Reelfoot Lake

*"And it shall come to pass;
wherever the river goes . . .
everything will live."*

(Ezekiel 47:9)

Reelfoot Lake

Oasis on the Mississippi

Jim W. Johnson

The University of Tennessee Press / Knoxville

Frontispiece: anthony heflin / Shutterstock.com.
All other images property of the author.

Library of Congress Cataloging-in-Publication Data

Names: Johnson, Jim W., 1940– author.

Title: Reelfoot lake : oasis on the Mississippi / Jim W. Johnson.

Description: First edition. | Knoxville : The University of Tennessee, 2023. | Includes
bibliographical references and index. | Summary: "This book explores the cultural and
environmental history of Tennessee's only natural lake, Reelfoot Lake. Jim W. Johnson, a
longtime west Tennessean and environmentalist, guides readers through the ancient and
recent history of the lake, provides a section for visitors, campers, and tourists interested
in visiting Reelfoot Lake, and discusses the past, present, and future conservation issues
facing this important body of water"— Provided by publisher.

Identifiers: LCCN 2022053789 (print) | LCCN 2022053790 (ebook) | ISBN 9781621907084
(paperback) | ISBN 9781621907091 (adobe pdf)

Subjects: LCSH: Lakes—Tennessee—Reelfoot Lake. | Natural history—Tennessee—Reelfoot
Lake. | Reelfoot Lake (Tenn.)

Classification: LCC QH105.T2 J64 2023 (print) | LCC QH105.T2 (ebook) | DDC 508.768/
1—dc23/eng/20221116

LC record available at https://lccn.loc.gov/2022053789
LC ebook record available at https://lccn.loc.gov/2022053790

Contents

Acknowledgments xi

Introduction 1

Chapter 1. The Creation of Reelfoot Lake 5

Chapter 2. Description of the Lake 15

Chapter 3. The Earliest Pioneers 25

Chapter 4. The Settling of Reelfoot Lake 33

Chapter 5. The Gumbooters 37

Chapter 6. The Life of a Kid on the Bayou 69

Chapter 7. The Natural Resource Managers 95

Chapter 8. Solitude, Changing Times, and Lore 105

Chapter 9. Rising Conflict 113

Chapter 10. The Mississippi Flyway 119

Chapter 11. Waterfowl Hunters, Duck Blinds, and Sportsmanship 135

Chapter 12. Exploring Reelfoot Lake with the Seasons of the Year 147

Chapter 13. Special Topics 183

Chapter 14. Auto Tours 197

Chapter 15. The Outlook 205

Appendix 1. Other Books about Reelfoot Lake 225

Appendix 2. Evolution of Oxbows and Maturity of the Floodplain 227

Appendix 3. Wildlands to Domestic Lands 229

Notes 231

Index 235

Illustrations

Figure 1. Reelfoot Lake.. color section

Figure 2. Historical Mississippi River channels every 500 years 11

Figure 3. A Sunday boat ride with granddad
with granddaughters on the Bayou du Chein 17

Figure 4. Profile of Reelfoot Lake; before and after
the earthquake 19

Figure 5. Cypress trees: a remnant of the former forest of 1811–12 20

Figure 6. Map of Reelfoot Lake 21

Figure 7. American lotus blooms... 22

Figure 8. An aerial view of Kentucky Bend............................... 23

Figure 9. "I found the Lake aglow with Yellow Lights
of the Great Lilies".. color section

Figure 10. A cabin near Samburg (1800s) tolerates flooding
by floating on logs... 34

Figure 11. "Ras" Johnson raised six happy Gumbooters
with wife Mildred 47

Figure 12. Floodwater was three feet higher than normal—
summer pool during local floods of 2011 and 2019.................... 57

Figure 13. "Boochie" Berry's set net; commercial fisherman,
and Reelfoot historian 60

Figure 14. Wendel Morris and Terry Pride brave the cold
to set trammel nets.. 62

Figure 15. Johnson family on the Bayou du Chein 72

Figure 16. Walnut Log boardwalks across the Bayou du Chein 75

Figure 17. Tennessee Academy of Science biological station 78

Figure 18. Keystone loggers: A mule team load of logs.................... 79

Figure 19. Barefoot boy and his dog on Bayou du Chein
at Walnut Log... 87

Figure 20. Baby alligator snapping turtle 88

Figure 21. Bennett Johnson cautiously puts a snapping turtle
in a fish live box 90

Figure 22. Black Bayou at sunset.............................. color section

Figure 23. Lake freeze-up and mallards on ice at my dock.............. 101

Figure 24. Infrared photograph of Reelfoot Lake............. color section

Figure 25. Foggy morn at the Ellington Center Boardwalk... color section

Figure 26. *U.S. Marshals* movie set; night chase for Wesley Snipes 106

Figure 27. *U.S. Marshals* movie set; Laker movie stars................... 107

Figure 28. Walnut Log or Ward's Lodge, 1908 116

Figure 29. Newspaper photograph circulated widely in the
Reelfoot Lake area concerning murder by the Night Riders 117

Figure 30. Cotton bolls in late September display
their soft, fluffy beauty 120

Figure 31. Gossiping snows, blues, and Canada geese
on Black Bayou Refuge... 121

Figure 32. Mississippi flyway map.. 122

Figure 33. Gordhead cranes return to Reelfoot Lake, 1990s............. 129

Figure 34. Redwing blackbird on its spring singing perch;
and male widow skimmer dragonfly................................... 130

Figure 35. Mud snake; harmless but vicious looking.................... 132

Figure 36. Tommy Lovell's duck blind and sunrise mallards 138

Figure 37. Elbert Spicer hunted ducks from the tip
of a cypress tree during the 1930s...................................... 140

Figure 38. Removing illegal duck blinds to restore order
and equitable hunting methods....................................... 145

Figure 39. Green tree frog hidden in lotus bloom............. color section

Figure 40. Baby painted turtle .... color section

Figure 41. Marsh birds on Reelfoot Lake WMA—Black-neck stilt
and least bittern.. 149

Figure 42. Muskrat sunning, Horse Island Canal............. color section

Figure 43. Gold finch on wild thistle, button bush,
and prothonotary warbler with green worm .... color section

Figure 44. Least bittern.. 151

Figure 45. Author's fresh catch: Two-pound crappie 151

Figure 46. Pileated woodpecker at Horse Island Canal 153

Figure 47. Invasive water willow 154

Figure 48. Juvenile bald eagle with parents 156

Figure 49. Juvenile eagle, close up . . . . color section

Figure 50. Tree swallows in a small cypress over water 157

Figure 51. Airpark campground trail photos 158

Figure 52. Forester terns, Upper Blue Basin 160

Figure 53. Bull frogs ready to spawn 161

Figure 54. Soft shell turtle pair on log . . . . color section

Figure 55. Sunrise and sunset are wonderful times
on the lake . . . . color section

Figure 56. Native wild hibiscus flower 166

Figure 57. A Mississippi River view from a dike
in late summer and fall 167

Figure 58. A Mississippi River sand beach, common
during late summer and fall 167

Figure 59. Bobwhite quail, once abundant in Lake County 169

Figure 60. Great blue heron tries to avoid spines of catfish . . color section

Figure 61. Our dog Wynette hiking a grassy island trail
and board walk 170

Figure 62. Egrets on stumps drowned by New Madrid earthquake 171

Figure 63. Joe Guinn stands among tall cypress knees 173

Figure 64. Fall on Fishgap Hill, Chickasaw Bluff 174

Figure 65. Snow geese, Black Bayou Refuge 175

Figure 66. November egrets go to roost . . . . color section

Figure 67a. White pelicans, Brewer's Bar 175

Figure 67b. White pelicans, Brewer's Bar . . . . color section

Figure 68. Bruce McQueen's old home place on Bayou du Chein 177

Figure 69. Mallards on ice waiting for thaw 178

Figure 70. Winter freeze-up at Gray's Camp 179

Figure 71. Sleepy bobcat 182

Figure 72. Fall pelicans raft in Open Lake 185

Figure 73. Spider-rigging is a favorite crappie fishing method
at Reelfoot Lake. 187

Figure 74. Mr. "B" runs trotlines out from Gooch's Landing,
Lower Blue Basin . . . . color section

Figure 75. Cottonmouth snake . . . . color section

Figure 76. Smooth back green snake on Spatterdock Pad 190

Figure 77. A tiny green cricket frog in duckweed . . . . color section

Figure 78. A bald eagle family in spring cypress 193

Figure 79. Ellington Center Boardwalk 194

Figure 80 A female redstart on equisetum during early spring 195

Figure 81. Winter day treat on Reelfoot trails and boardwalks 198

Figure 82a. Barrowed owl chick 199

Figure 82b. Cardinal flower 199

Figure 83. Ancient bald cypress trees . . . . color section

Figure 84. Fishermen wading among giant cypress trees 201

Figure 85. Mississippi River levee 209

Figure 86. Aerial view of the Mississippi River
and Reelfoot Lake 218

Acknowledgments

To Dr. James (Jim) Byford, University of Tennessee professor emeritus, I owe much; he has always been available to generously offer advice and encouragement for many natural resource conservation endeavors during my career, especially those that concern the promotion and preservation of wildlife and native wetlands. Dr. Byford is the author of *Close to the Land* and has encouraged the writing of this book, reviewed it, and encouraged its publication. For his many suggestions, edits, and encouragement, I am extremely grateful.

To Stephen Lyn Bales, senior naturalist at the Ijams Nature Center in Knoxville, Tennessee, and author of *Natural Histories*, *Ghost Birds*, and *Ephemeral by Nature*. His attention and clear insight as to the arrangements of chapters and topics within the manuscript was a blessing, something I wrestled with throughout writing this work. His review was also inspiring, and I am extremely thankful for his critical analysis and editorial contributions.

To George Grider, who recently published a wonderful book, *Crossing Borders,* with John Wilson Spence, I am grateful for his early suggestions on chapter arrangements.

To retired teacher Betty Krone, my dear mother-in law, who pointed out numerous grammar glitches and patiently read to me when I was flat on my back in convalescence a season or two ago. Her positive outlook about the book and everything else in life is truly inspiring.

To Sharon Cunningham, a native of Reelfoot Lake, born and raised in Samburg, Tennessee, I am most grateful. Sharon knows as much as anyone about Reelfoot Lake and its people; she was an editor, among other duties, at Dixie Gun Works for many years. Sharon generously volunteered to preview an early draft of the manuscript and encouraged completion for publication.

To my wife, Kathy, who has patiently tolerated my extra share of computer time for this work, and for coming to my rescue many times to bail me out of lock-down, I am greatly appreciative. I am also thankful for her instructions to help improve my computer skills.

Introduction

There it lay: twenty-five-thousand acres of shimmering water and emerald green. It is arguably the most unique and biologically rich wetland nature ever created! With its four large basins and cypress groves, haunting cypress swamps and lily marshes more than two-hundred years old, it is an aesthetic mix of astounding natural beauty—and happens to be along the largest migratory flyway in the world. Rich in West Tennessee natural and human history, this verdant wetland, surfeit with ecological diversity, is the destination for those not content without a frequent and generous dose of nature. You'll hear such testimony from the quarter-million outdoor users and visitors that annually come here. Another thing: accommodations are plentiful and local folks as friendly and helpful as you will find. There is only one stretch of native wetlands along the Mississippi River between Lake Itasca, Minnesota, and the marshes of Louisiana that fits this description so well—Reelfoot Lake.

I was fortunate to be born and raised there, and to this day I can say there was no better life for a kid who loved nature and the great outdoors. There's simply nothing like it in Tennessee, or any other place I know within a day's drive from West Tennessee. In spite of its reputation as the state's region for abundant wetlands, without Reelfoot Lake, West Tennessee would be virtually without a large *native wetland*—especially one with such unique creation. As a kid guide for fishermen, sightseers, and sometimes scientists and students who came to investigate the lake's ecosystem, or just to collect specimens, I'd never heard the words "wetland ecology" or "food chains." But I had a sense about those terms at this early age. It was easy duty because learning more about the nature of the lake was my favorite thing to do. Although we had very little written information on the history of the lake, we were surrounded by old timers (and our own families) who loved to tell what they knew about the lake. Of course, that information was included in my guide service.

One thing I learned was that an exceptional guide—in rain, shine, sleet, or snow—could make a huge difference in the ultimate success of a visitor's trip, especially for those with limited time. So that might help answer a familiar question: "How is the best way to enjoy and get the most from a visit to a natural area like Reelfoot Lake?" You might say, "Get an experienced guide." I agree, especially if you are unfamiliar with the area. But it is not always practical to have your own personal warm-blooded woodsman for a guide.

Reelfoot Lake was never far from my thoughts. It was long on my mind after leaving my home at the lake to answer the call of other goals when the idea for a guide book about the lake really began to gel. That was when I became the director and programmer for the National Wildlife Federation's first and second national environmental youth camp at Camp Energy, Land Between the Lakes. Here's the story.

It was, of course, a primitive tent camp—outside, in the woods, exactly where I loved to be. Odd that it might remind me of Reelfoot Lake, but it was like guiding a camp of youngsters; as a kid guide, I tried to provide a an enjoyable, meaningful, and rewarding outdoor experience for my clients—exactly the objective we intended for these 240 young outdoor adventurers, all running on high-octane energy. Appropriately enough it was called "Camp Energy."

It could not have gone better. By this time, I had learned something about the words and science of "ecology" and "food chains." So, with a little practical experience and a little science, writing a program for our wonderful camp guests was not too difficult. It dove into things of nature and explored the history of former travelers, their ancestors, and Native Americans who might have passed along the same trails the young camper traveled. Our counselors followed that program with great success. With this commitment—along with a colorful hands-on experience—the woods came alive for these young outdoor adventurers—and our mission was accomplished. Some 240 happy campers returned home to tell their stories.

With Reelfoot Lake never far from my thoughts, I began to wonder: why hasn't someone written a program like this for visitors to the lake? No, the experience would not be the same as having a warm-blooded woodsman at your side, but with a good virtual guide book, it could be close. This way, visitors could take their own personal guide with them.

That idea was actually the beginning of this book. But I was a long way from Reelfoot Lake and the idea began to fade as the years went by.

Meanwhile, the importance of native wetlands like Reelfoot to the nation was made more real over the next two years with the Federation when I was transferred to the big city—Washington, DC. There could be no better reminder of this importance than to consider those obliged to spend most of their lives in the city; there're no real substitutes for a large natural area. City parks are about the best such cities can do. While Washington, DC, like many cities, is a beautiful setting, with its cherry blossom springs, for people without several days of free time, the great outdoors was almost out of reach. While city parks are tremendously important, access to native wild lands (or just knowing they are within reach) is inherently comforting. The experience was just another reminder that we have a great need for well-managed public natural areas like Reelfoot Lake.

As I neared the end of my career and moved back to Reelfoot, I found that while a number of books had been written about the lake, a comprehensive guide book for visitors was yet to be written. I decided it was a good time to give it a try. A daunting venture? Yes, it was. But after a couple of years and some interesting research, the last chapter of the draft manuscript was finally ready for edit in late winter of 2021. Little is spared about the lake's creation, its rather unique society and way of life, its natural history, or the virtual guided tours. It is also partly a story about the ecological ties of Reelfoot Lake to the great Mississippi River, and its floodplain uses, benefits, and our influence on its natural ecology. Along the way, you'll be introduced to some of the people—the Lakers, the friendly and helpful community often mentioned. One of the first things you will note is that, like me, they have an uncommon passion for Reelfoot Lake; the mere mention that they might uproot and leave is a loathsome subject. One who has expressed that feeling so adamantly is a life-long resident, park guide, hunter, fisherman, and historian of the lake—Mr. Lexie Leonard. In his book *Reelfoot Lake Treasures,* he speaks most eloquently about the lake on our behalf, noting: "I have to be very careful not to worship the creation more than the Creator."

You will understand Mr. Lexie's compassion for Reelfoot Lake after a single visit. Whether the reader is a naturalist, a birder, hunter or fisherman, or one who simply enjoys scenic views, I trust this book will be a rewarding and unexpected adventure.

Chapter 1

The Creation of Reelfoot Lake

The Earthquake and the Mississippi River Boatmen

The night of February 6, 1812, was not unusual; a calm moonlit night, cool and clear, a light breeze carried the fresh odor of fish as it drifted along with the meandering river. Vincent Nolte and his friend Hollander had tied up their flatboat at the little river port town of New Madrid on the west side of the river. About a dozen or so other boats from Shipping-port, an Ohio River port at Louisville, Kentucky, had moored with them. A small campfire on the sandy shore flickered dimly on the faces of a few worn riverboat men who'd stayed up a little later than usual to enjoy the warmth of the last burning coals before sack-time. It had been a hard day. Maneuvering a flatboat all day required a lot of sweat and manpower, es-pecially when fighting the strong currents of this river.

Conversations had finally run their course, and half of the late-evening crew had drifted into the night toward the gangplank. Time to kick out the fire. There was little else to be mindful of but the powerful pulse of the Mississippi and the noisy slurping of hungry eddies that swirled down-stream one behind the other. The river was on a rise.

Hollander had gone to bed two hours before the riverboat men, but Nolte was still up at midnight painting a caricature of President James Madison. "I had just given the last touches [to the art work]," he said, "when there came a frightful crash, like a sudden explosion of artillery . . . countless flashes of lighting; the Mississippi foamed up like a boiling caldron, and the stream flowed rushing back, while the forest trees, near which we lay, came cracking and thundering down."[1]

How could they know that they had landed at the epicenter of one of the greatest earthquakes in North American history—the New Madrid Earthquake? Self-preservation demanded they escape—somehow. But where would they go? No one could hide from such a horrendous and omnipresent disturbance. Their overnight neighbors from Shippingport, desperate to do something, had loosened the tackle of their mooring in an attempt to escape the uncertain turbulence; the rage of the river was nothing compared to the cavernous fissures of the howling ground, distorted, and spewing sulfur ash. Now they struggled to gain control of their boats. But the powerful turbulence of the current forced them farther and farther away—upstream, as it were, against their will, and against the ordinary nature of the river—into an unfamiliar darkness.

Nolte and Hollander would not hear from them again. At sunrise, they faced the melancholy truth: they were alone; the town of New Madrid had been demolished, the fate of its two-hundred inhabitants yet unknown. They knew of no reported deaths; that is, other than that of an Indian hunting party they'd heard about from an earlier tectonic event. Seven Indians in canoes nearby, the word was, had perished in the roiling caldron of the river, their fate known only because one of them escaped to tell the story.

A few anxious weeks passed. Then came a report from Indian country across the river; it seemed to confirm that a huge sink had developed in the earth between the river and Chickasaw Bluff some seven or eight miles east, and the river had filled it. Some claimed it was as much as six miles wide and twenty miles long. But maybe twice that. At any rate, there was little doubt that a new lake of some exception had been formed. Little else was known about it for many months. Word eventually trickled out that it was true—the lake was, indeed, large and remarkable in its beauty and splendor. Myths and fables to explain how the lake was formed quickly spread as local lore, much of which continues to this day. But most were content to give credit to the earthquake for what would become widely known as Reelfoot Lake, "The Earthquake Lake," "The Swamp Country"—a paradisiacal wetland destined to be my future home.

The New Madrid earthquake was the second of two unique events in the creation of Reelfoot Lake. The first of five tectonic events occurred at 2:00 a.m. on December 16, 1811. Little Prairie, thirty miles downstream of New Madrid, was destroyed by liquefaction. The last of these was reported at 3:15 p.m. on February 7, 1812, which registered (depending on the authors of the report) 7.5 m (or 8.5) on the Richter scale.[2]

Shifting land and water, great upheavals of earth, sulfur smoke, and chunks of coal, and a seething sunken forest took more than two months before the earthquake settled; and a few landers survived to tell their story. Eliza Bryan, who lived on the west side of the river, gives us one of the most vivid accounts of the earthquake in a letter of 1826 to her brother.

Her account is often found printed on table mats at local restaurants. Here is an extract from her letter: "The moon was shining brilliantly," she reports, "but sulphurous vapor caused the earth to be wrapped in absolute darkness. The wailing inhabitants, the stampede of fowls and beasts, the noise of falling timber, the roaring of the Mississippi—the current of which was retrograde for a few minutes—formed a scene too appalling to conceive of."

When the earthquake ended, a phenomenon of great natural beauty, an aquatic garden with a large lake in which only the tops of trees were visible, and bordering wetlands that covered some thirty-thousand acres, lay quietly over the floodplain. And this uncommon inheritance of natural resource wealth and pleasure was passed on to us as caretakers to enjoy. The awesome power of nature always precedes great creations such as rivers, lakes, canyons, deserts, and mountains, and they are blessed with breathtaking beauty, fauna and flora, depending on their geography. Reelfoot is a shining example.

So, what makes Reelfoot Lake so unique is the influence of two independent natural events in its creation—a river that had long since created a series of oxbow lakes and associated wetlands; an earthquake that almost instantly sank and dammed up the river's work, which was then filled by the river and runoff over a period of time.

A river, you say? Yes. So the popular notion that an earthquake created the lake is only partly true. The Mississippi also had much to do with the creation of Reelfoot Lake. *While it can be said truly that nearly all native wetlands in the Mississippi River watershed are created by rivers, Reelfoot Lake is the odd exception: The river is the ecological mother, but the geological form of the lake basin was created by the earthquake.*

Another lake like Reelfoot is simply unknown. Reelfoot Lake is the largest, most biologically diverse, and arguably, the most aesthetically beautiful natural lake in the world. There are some small lakes on the west side of the Mississippi, like Brandywine Shute, Wapanocca Lake, Horseshoe Lake, Chicot and Midway Lake in Arkansas, and several others in the Mississippi River drainage. Many have the characteristics of Reelfoot Lake. The signature of living cypress standing in open water are found in some—but these may or may not have been created by the earthquake, and none is as large as Reelfoot.

Old river scars and oxbows, like those in Reelfoot before the earthquake, may dry up and sprout trees before they are re-flooded. But the ecology of these lakes is nowhere near the complexity of Reelfoot Lake, for it is extremely fertile, carrying about six times the poundage of fish per acre than any lake in Tennessee. It is home to a resident population of American bald eagles, giant Canada geese, amphibians and reptiles, and an enormous population of resident songbirds. It is indeed a unique native wetland.

Because the relationship between the Mississippi River and Reelfoot Lake is so intimate, let's briefly explore this relationship. Understanding it will enhance a visit to Reelfoot immeasurably. Natural wetlands like oxbow lakes are not only created by native rivers, they remain intimately tied to them throughout their existence, refurbishing the wetland with water, nutrients, and aquatic life. Like a living organism, native rivers constantly meander about their wide floodplains, adjusting to be more efficient, defying manmade interferences, creating new and doing away with the old, refreshing themselves in every way—and they have no respect for people. Without the natural flow of a river, we would forfeit nearly all of our native wetlands. That is to say, artificial "rivers" produce artificial wetlands; these are substitutes, low-value wetlands (if any net worth), and not ecologically self-sustainable like native rivers. And most are to our detriment, as they replace and function without any of the benefits of native streams and wetlands.

Native rivers are like living things, always changing, adapting, and not inclined to stay in the same place. Channels of the Mississippi have been on both sides of the floodplain many times over the millennia (Figure 2), sometimes suddenly, as in avulsion. But our modern-day priorities often do not allow these adjustments to take place. Dams, levees, highways, and other arbitrary adjustments are incompatible, divert, and destroy the purpose of the river. The Mississippi is the mightiest river in the Lower Forty-Eight.

Mark Twain said of the Mississippi River that there were fifty-four sub-rivers, and hundreds of other tributaries, and that it is the longest river in the world—4,300 miles of natural resources within its corridors. That is, before the Industrial Revolution had occurred within something like the last hundred years. It is also considered "the crookedest river in the world, since in one part of its journey it uses up one thousand three hundred miles (1,300) to cover the same ground that the crow would fly over in six hundred and seventy five (675)."[5]

So about half of the river miles are busy scouting the floodplain to create the most efficient drainage; in the process, creating ecologically natural wetlands like Reelfoot Lake, one of the most biologically diverse and scenic ecosystems in the world. So, Reelfoot Lake is actually a component of the Mississippi River. Imagine the enormous natural resource potential of reasonably natural rivers, especially one that runs 4,300 miles, which had something like twenty-five million acres of native wetlands!

Why is all of this important? For one thing, it creates natural wetlands like Reelfoot Lake. But the river also does a far better job than we humans when it comes to accomplishing its purpose: that is, to carry runoff, to prevent catastrophic flooding, and to harmonize its ecosystem. That is mainly because we usually have tunnel vision about how to use this great natural resource; it cannot be a utilitarian natural resource, only self-serving or for a single purpose. So that brings us back to the Mississippi: How did we decide that the priority use of this river was for transportation and agriculture and not for ecological diversity, for fish and wildlife, the fastest hardwood growth on earth, floods and wind control, soil conservation, ground water resources, or to feed the oceans? Was it the best use of natural resources? The best net value?

Well, no, not any of that. The river channel was altered for a misused word called *progress*. It is an ill-defined term: progress for whom? Is the river's value perfunctory for natural resources when the term is applied? Progress seems meaningful only to the giants of industrialization. Use of the river should require a broader view. How about considering the economic value of natural resources alone—lumber production, ground water recharge, wind protection, and soil and chemical filtration, fish and wildlife, outdoor recreation, the creation and restoration of healthy wetlands, and so on. And only of late have we learned that diverse natural ecosystems harbor fewer, not more, diseases. That is because biodiversity dilutes pathogens like viruses. Biodiversity provides a natural shield that mitigates fallout from these pathogens.[4] Somehow, the net worth of these values was ignored in the economic equation. Economists also ignored that natural resources are self-sustaining; all they really require is protection and equitable allocation of their surpluses. All of it is a sustainable and useful product, not to mention intangible, non-consuming value (a family outing, scenic views, hikes along foot trails, education, bird sanctuaries, etc.) that cannot be expressed in money alone.

So, that brings us back to Reelfoot Lake: Now that we have a hint about the dynamics and use of a river, we can follow the unusual partnership among Reelfoot Lake, the Mississippi River, and the New Madrid Earthquake. Once the earthquake sank part of the wetlands outlined as Reelfoot Lake, the Mississippi, as told many times, flowed upstream and returned temporarily to its old path through these wetlands. We don't know how much of the lake was filled at this time but some witnesses living in Madrid Bend (a rare few lived along the Mississippi at the time of the earthquake) say it took months to fill, contrary to "old wives tales" that the lake was filled instantly when the Mississippi was reported to flow retrograde.[6] But when it did fill, it inundated mega-tons of living plants. It must have been a steamy caldron the following summer, a rich soup of green vegetation. Along with its nutrient-rich soil, tons of organic material lay seething in the hot days of summer. It took a while for all of these chemical and biological changes to establish some kind of biological equilibrium—very likely an erratic process with some fish die-offs and some flourishing during the first year or two. After that, the lake undoubtedly settled down and flourished with native fish and wildlife.

Can you imagine the results of such an event? It must have been awesome! My grandfather had some amazing stories about life at the lake some fifty years after the earthquake. The lake was alive, he said, with breathtaking freshness, and it swarmed with bird life, fish, and creatures of all kinds. Probably, nothing like it had been created in recorded history. Eventually, the river flooded and scoured the land in some places and deposited sediments in others. Finally, it ended up like something near the present pool and configuration. You will see footprints of this history by the living cypress trees in the open lake as soon as you arrive.

Try to imagine the frightful event that created Kentucky Bend. The disturbance must have petrified creatures for miles. It happens when the river choses to change its course. A sudden switch in the Mississippi River channel, called an *avulsion,* is nothing to wink at; it can take land from one owner and add to another. It's awesome. It has happened many times over the millennia. But the switch to create Kentucky Bend is the most extraordinary one in our stretch of the river. It occurred, perhaps five-hundred years ago when the river was at flood stage; in the blink of an eye, the river exploded. It would have been something like rolling thunder—wham! And then the earth rumbled like an earthquake for hours. A storm of rushing water jumped its bank a mile or more upriver from where Reelfoot now lies, and headed west. Like a giant middle buster plow furrowing the ground, a new channel was formed.

Within the span of perhaps a day, the river had switched its location and went in a loop about twenty miles toward the small town of New Madrid, Missouri, that stands today. Founded by the French and then the Spanish in 1718, it probably did not exist at the time. At New Madrid, the new channel of the river made a wide bend back toward the east to unite

Figure 2. Historical Mississippi river channels
every 500 years.

with its original main channel. The entire loop created Kentucky Bend (known also as the New Madrid or Bessy Bend). This change in the river set the stage for future Reelfoot Lake.

The disturbance caused the river channel to vacate the eastern flood-plains, which left a thirty-mile region of wetlands close to the Chickasaw Bluff, parallel to the river (where Reelfoot Lake now lies). The low-lying wetlands would become the bed, an ecological foundation for future Reel-foot Lake. As you will soon see, the final part of the creation of Reelfoot Lake will be the result of the 1811–1812 earthquake.

Stages in the Creation of Reelfoot Lake

Let's consider the two major events that created Reelfoot Lake. *First by a river:* over millennia, the river left a series of oxbows, swamps, and marshes, which formed the foundation of the lake. *Second, by an earth-quake:* The second event was the sudden impact of the New Madrid earth-quake. The phenomenon and its sequence are most important to know, because both are major works of the earth's master contractor—nature. Since nature does nothing foolish, either event would have left something extraordinary and of special ecological benefit. We could expect no less than the unusual and magnificent ecosystem of Reelfoot Lake.

A unique wetland formed. So a large, mature wetland, itself teeming with biological communities, suddenly sank and filled with water, be-coming a fresh new lake that might have been as much as forty-thousand acres. It is difficult for us to imagine how extraordinary an event like this can be given the abundance and tremendous vigor of fish and wildlife activity.

Contrast manmade or artificial lakes with natural lakes like Reelfoot. Manmade lakes often attempt to duplicate the creation of natural lakes with artificial dams for various benefits, usually hydroelectric with out-door recreation benefits. But manmade lakes are, at best, mere shadows to the ecosystems and benefits of a natural lake, although they can pro-vide enormous benefits in storing water, providing hydroelectric power, hosting outdoor recreation, and prompting a surge real estate values. At the same time, however, large manmade reservoirs usually sacrifice the ecology of the river that feeds them.

Barkley Lake is one such manmade lake. This huge reservoir in Ken-tucky and Tennessee is on the Cumberland River; I worked on several proj-ects during its construction and was there when it was flooded during the 1960s. But the reservoir of Barkley Lake was cleared and grubbed; tons of green trees and other organic matter were not there to provide the extraordinary nutrient richness that existed at Reelfoot. So, I have no doubt that accounts of the extraordinary abundance of Reelfoot Lake fish and wildlife told by my grandfathers and the old timers were not exaggerated one iota. In fact, the lake was so bountiful with nature, they were often at a loss for words to describe it.

Yet, by this time, Barkley Lake was ready to be flooded. Low shrubs and weeds had grown up since the reservoir was cleared, providing an enormous source of seeds, insects, and other food sources for fish and other wildlife. Once the lake was flooded, the response by waterfowl and fish for about the first three to five years was so tremendous as to defy description. It reminded me of the extreme drought years at Reelfoot, when much of lake's bottom had been exposed and dried; the lake's ecological condition was dramatically refreshed with vigorous growth of fish and wildlife populations.

While Barkley Lake could never be as biologically diverse and nutrient-rich as Reelfoot Lake, it demonstrated valuable lessons about the creation and future of natural lakes—even Reelfoot Lake. One is that the early development of oxbow lakes (many old ones are beneath the waters of Reelfoot Lake) is very similar to the first two or three years after Barkley Lake was flooded: (1) The treeless reservoir grows up in shrubs and weeds during the dry summer before being flooded, which provides a tremendous source of food for fish and wildlife. Consequently, the variety and population levels of fish and wildlife for that year increase dramatically. (2) It demonstrates the annual effects of natural seasonal drying and flooding. That flooding and drying cycle occurs in all native rivers and wetlands: their floodplains dry during dry seasons, and flood during wet seasons. This climatic cycle is the driving force that sustains the biological health of all native rivers and their wetlands. Managers often try to duplicate this phenomenon to improve the health and vigor of artificial lakes or artificially controlled natural lakes.

Chapter 2

Description of the Lake

One thing to clarify up front is that Reelfoot Lake is a rare and fragile piece of public real estate—one major reason being that, other than its unique creation, the lake was born a natural wetland, an endangered habitat type within the Mississippi River regions. Native wetlands are rapidly disappearing, and few new ones are being created. Such a conception requires a strong bond between native rivers and native wetlands, as native rivers create nearly all native wetlands. And while our former river floodplains are abundant with wetlands, they are generally modern wetlands, that is, artificial ones. In contrast to native wetlands like Reelfoot, these modified rivers are considerably inferior and destructive compared to wetlands created by free-flowing, natural rivers. And artificial rivers defy natural river ecosystems. Of the ten major rivers (the Mississippi is included) of the world today, only the Amazon remains a free-flowing river. This means that the Mississippi (and nearly all of its tributaries) are no longer classified as native rivers and cannot create native wetlands, although some may temporarily exhibit natural characteristics. All of the tributary rivers in this region, which have been artificially modified—converted to ditches or artificial channels, create *artificial wetlands*.

How does this work?

Native alluvial rivers like the Mississippi are dynamic and free to move about their wide floodplains as the need arises to more efficiently carry runoff. In the process of change, the river channel moves from one place in the floodplain to another; oxbows and other wetlands are created—oxbows like those that lay the foundation of Reelfoot Lake before the earthquake. Oxbows grow old; these fill with sediments, and become marshes and swamps. Eventually, the wetland is artificially drained for other purposes, or through natural succession, and becomes a bottomland hardwood forest. Few, if any, new oxbows or other native wetlands are created. Thus, within the floodplains of the Mississippi River and her tributaries, we have lost perhaps 90 percent of the native wetlands present during the first decades of the 1900s. Think of the loss from those voids to the natural world. Can they be replaced? It leaves no doubt that the preservation and ecological health of Reelfoot Lake should be one of our highest natural resource conservation priorities.

The earthquake changed everything. Before the New Madrid earthquake, the low floodplain where Reelfoot now lies was a series of old oxbows, marshes, cypress swamps, and a hardwood forest that surrounded this complex of wetlands. It extended from about Hickman, Kentucky, south to the Obion River. The Mississippi was a free-flowing river back then.

Immediately after the earthquake, in 1811–1812, the uplift that created a natural dam below the lake backed up thirty- to forty-thousand acres of water that lay parallel with the Mississippi River. Today, Reelfoot Lake public lands in Tennessee contain some 25,579: more than fifteen-thousand acres are in navigable open lake and five thousand in cypress swamps and upland hardwoods. The newly established Black Bayou Refuge and buffer zones around the lake contribute an additional four-thousand acres of managed wetlands. The open basin is about seven miles as the crow flies from the north end of Upper Blue Basin south to Lower Blue Basin at Champey Pocket. Forty miles of highways circumvent these wetlands, with many side roads to the lake. The lake has four major basins: Upper Blue Basin, Buck Basin, Middle Basin, and Lower Blue Basin. Reelfoot is shallow with an average depth of 5.6 feet. The maximum depth of Upper Blue is about twelve feet; the maximum depth of Lower Blue is eighteen to twenty feet.

The deepest part of Reelfoot Lake, the Washout is a small appendix-like basin that extends beyond at the south shores of the lake. Created more than a half century after the New Madrid earthquake, it is still a part of Reelfoot Lake. The Washout has sandy beaches, few or no stumps, and a maximum depth of forty feet (once known to be eighty feet deep); it was created during the raging floods of the 1880s by the Mississippi River, mainly by the floods of 1883–1884. These floods broke through the natural dam, now Highway 21, spilled and soil-gouged out the place we now know as the Washout.

There were no government levees back then. So, without the levees, the river leisurely included Reelfoot during floods as an auxiliary overflow channel. When this happened, one could take a boat ride from the Chickasaw Bluff and go east nearly sixty miles to the high grounds at

Paragould, Arkansas.[1] Since most of the land was forested, very little soil erosion occurred. But, once the land was cleared, the Washout and much of the land below its drainage eroded away. Sunkist Beach benefited from a considerable load of sand during these floods, compliments of the river. This beach and Magnolia Beach were very lively concessions when I was a kid. Today, it is still popular for water skiing and some beach activity.

Reelfoot, by its very nature, is a scenic and placid lake. Quiet cruises on the lake with or without a guide are greatly understated. Watercrafts like canoes, kayaks, stump-jumpers using oars, paddles or push-poles are excellent for low-impact ways to enjoy the lake—and for the very pleasure of using the primitive equipment. Try rowing a Reelfoot stump-jumper with bow-facing oar locks for the first time; some find it a bit tricky without practice. Trolling motors, or other quiet motors on pontoons, canoes, and other watercraft also work well. "Quiet" is the key word; leave frolic where frolic is sanctioned.

Reelfoot has plenty of opportunities for quiet boat trips. Many miles of canals, ditches, and natural waterways wend and wind through the marshes, swamps, and basins. There is a basic difference between these waterways: *Canals* are manmade for navigation. *Ditches* are manmade for distributing water. *Natural waterways* are drains to carry runoff and for fish and wildlife pathways to and from brooding grounds. Knowing the difference is important for managers and users, as the reckless alteration of natural streams and wetlands is the primary culprit for the greatest loss of national wetlands. Many natural rivers and smaller streams have been arbitrarily altered and turned into ditches and canals. When such

Figure 3. A Sunday boat ride with granddad
and granddaughters on the Bayou du Chein.

conversions are pressed upon these waterways, their ecosystems collapse and lose the characteristics of a natural stream. But canals and other boat channels found at Reelfoot have been constructed with forethought and are compatible with nature and outdoor use.

Earthquake Subsidence

What *part* and *how much* of the lake sank? The New Madrid earthquake did not appear to sink the entire lake as we know it today. In fact, it might have taken six-thousand years of geological change for the basic formation of Reelfoot Lake, although there is no doubt that the earthquakes of 1811–1812 had a great deal to do with it. How was the dam created that held the waters of the lake? Similar queries were long open to conjecture when I was a kid. It makes sense that a depression in one part of the earth could cause a concomitant rise in the earth nearby, and that created the dam, but that is far too simple for a geological answer. However, since the days of lively speculation, geologists have concluded that this is exactly what happened: the subsidence at Lower Blue Basin and the uplift below it created a dam long enough to cross the entire eastern floodplain and the river channel, which backed up the river and contributed to the flooded lake. Sediment deposition from frequent flooding by the Mississippi River (like that of an oxbow outlet) no doubt added to the height and depth of the earthen barrier.

We have plenty of evidence from tree snags (stumps) in the lake to give us clues about the filling and botanical status of the lake before and after the earthquake. As a kid living here, I was inquisitive about these indicators. So some novice scientific work was in order when I later returned to Reelfoot Lake: samples based on the depth of trees or snags beneath the lake surface indicate that a major portion of Lower Blue Basin sank ten or twelve feet at some point of time, but none, other than ordinary river scars and oxbows where the river once flowed, is evident in Upper Blue Basin. Consider this: the entire length of the former wetland, now covered by the lake, was one marshy wetland. Like the level of water in the lake today, the gradient of the floodplain in the cypress trees swamps, and standing water was within a few inches of the same elevation throughout the lake.

All I needed to do was measure the depth of the tree roots along the old oxbows and consider the average fall of the floodplain; ordinarily, the tree roots would be almost the same depth in both upper and lower basins. However, they were not: Lower Blue Basin tree roots were eleven to twelve feet deeper than those in Upper Blue Basin! With no other apparent reason for the subsidence, the cause points to the New Madrid earthquake of 1811–1812. I will spare the reader a rather lengthy explanation for these calculations, but it seems that the earthquake sank a major portion of Lower Blue Basin by about 11.2 feet.

We also know the earth is some ten to twelve feet higher than normal ground immediately downstream of the lake, which creates a natural dam. Much of it includes the Tiptonville Dome—considered the highest natural ground in Lake County. With the earthquake dam in place, the Mississippi River at flood stage and flowing retrograde, river water could have readily spilled into old river channels upstream, old channels that once drained the sunken wetlands. Within a relatively short time: Presto!—Reelfoot Lake was born.

The new lake, with its family of oxbows and old river meanders, stretched about thirty miles beside the great river, from Hickman, Kentucky, south to the Obion River. It must have been a biological wonderland! It was arguably the largest and most ecologically complex oxbow wetland of all the lakes along the two-thousand-mile stretch of the Mississippi from Canada to the Gulf of Mexico. Amazingly, it probably settled down and took its final form in about two months. We should take notice of how awesome the extremes of nature can be in power and consequence. For what we might consider catastrophic events, the enormous benefits of the aftermath are nearly always more than our assessments of the losses.

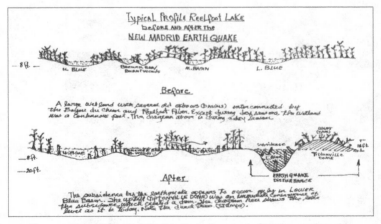

Figure 4. A profile of Reelfoot Lake; before and after the earthquake.

Oxbows, Snags, and "Mine Fields"

Beneath the rich waters of the lake are many old river channel scars, marshes, swamps, and former forests where the river had been centuries before the earthquake—decaying vegetation and dead animals; compost, you could say, by the metric tons. It is the main reason Reelfoot is extraordinarily nutrient-rich. The old channel scars are evidence of oxbows left in various stages of maturity caused by avulsion, a sudden switch in the river channel. Evidence of younger oxbows is in the open lake where stumps and surviving trees are absent, usually the deepest water. Trees are sometimes alive and struggling throughout the lake, but evidence of the former forest is mostly dead snags or stumps. Before being flooded, these same trees were alive and well. Other open water free of stumps is in the old channels of creeks and rivers partially inundated at the earthquake event. Reelfoot Creek (River) and the Bayou du Chein are examples. Knowing the trails through these stump fields makes navigation much easier. In summary, this is a "personality" sketch of phenomenal Reelfoot Lake.

As one might expect by now, oxbows (small natural river lakes) evolve, going through several stages as they age and mature. Such is the evolution of Reelfoot Lake. First we see an abandoned river channel; after a few decades, it matures as a small natural lake. Open waters fill with sediments and become cypress swamps and marshes. Finally, these wetlands fill and become a hardwood forest. Forests found in the floodplain are usually the last stages of former oxbows.

Figure 5. Cypress Trees: a remnant
of the Former Forest of 1811–1812.

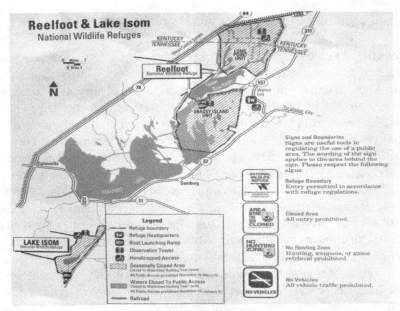

Figure 6. Map of Reelfoot Lake.

Snags or stumps? I call them "mine fields." Left to be inundated when the lake filled, most of the dead trees now appear to be "stumps"—really, snags, the tops of which were dead trees broken off over time. Some snags still stand well above water—they too will break off and become stumps. Today, most of the lake is replete with stump fields, usually unseen, but only a few inches beneath the lake's surface at summer pool. Going through a stump field can scare those unfamiliar with Reelfoot's stump hazards. But after a while it becomes easier and the boater pretty much ignores the rough treatment.

Though worrisome for navigation, stumps play an important role; they (a) are habitats for minnows and aquatic insects—fish food that attracts big fish; (b) good fishing spots; (c) act as "speed bumps" (speediness is not a compliment at Reelfoot Lake); and (d) are strategic fishing stands for egrets, herons, and other predator birds. So, if you notice a "ski-daddy" zipping across the lake as if it were his ocean, he either knows the stump-free zones or will likely regret his ignorance.

Ecology

The study of ecology (hardly known to Americans before the 1970s) was credited to Ernest Haeckel. But Alexander von Humboldt, a naturalist, who knew and wrote more about natural resources in the mid-1800s than anyone, including the evolutionist Charles Darwin. Reelfoot Lake is an excellent opportunity to think about the ecology of nature. Here, natural changes generally occur in an orderly fashion, as in the typical maturity of natural lakes: abandoned river segments become small lakes; lakes become marshes; marshes become cypress swamps, and on to bottomland forest.

These stages are noticeable from one end of the lake to the other at Reelfoot—a masterpiece of ecological expression, with nature changing a little with each season and each year. Reelfoot demonstrates the verity that nature abhors stability. Nature is never static; it is always in a state of flux, adapting to the climate, civilization, fire, storms, floods, and time. Every stage of change benefits certain species and disadvantages others, but the entire ecosystem follows a plan. As decades go by, wetlands evolve from an aquatic environment to an upland environment, aquatic plants and animals eventually evolve to upland plants and animals. All of these stages are in progress at Reelfoot Lake.

Adaption of Live Cypress Trees in the Open Lake

Are the cypress trees mid-lake living or dying? Living bald cypress trees are scattered throughout the open water of the lake. But trees do not sprout and survive in standing water—they die. Cypress trees are

Figure 7. American lotus blooms.

extremely resilient; these deciduous conifers have such great tenacity for life that they persist with all their might through storms, fires, and floods. Cypress swamps evolve from seasonal drying and flooding of the floodplain. So it is a matter of timing for cypress swamps to get a foothold; when their seeds sprout, the sapling must grow enough to keep its head above water through the following growing season—or perish. Once established, however, the floodplains must dry long enough during summer for the roots of the trees to take up nutrients—or they drown. Some trees, such as willows, are tolerant of a certain degree of flooding during the growing season, but cypress trees are extremely tolerant. Their roots need to be exposed to sunlight and air only one season every ten years or so, which is one good reason not to complain too much about dry years. Dry seasons and flooding keep wetlands healthy, wetlands like Reelfoot Lake.

Most of the living trees and stumps we see out on the lake were there in 1811–1812. Lucky for those living cypress trees on the lake today, they were tall enough when the lake first flooded to keep their heads above floodwater. Others were not so lucky. Trees like sweetgum, cottonwood, water maple, willows, and so forth at the water's edge, and most of the yellow poplar, sugar maple, hickory, oak, and others that require higher ground and drier root zones did not survive the flooding, as evidenced by the huge stump fields throughout the lake today. The cypress trees still living have slowly succumbed to standing water for more than two-hundred years. A few have sprouted on stumps or logs, but they, too, are likely to be victims of too much water on their roots. While they precariously survive, trees in the open lake die year after year.

Figure 8. An aerial view of Kentucky Bend (Reelfoot Lake visible at lower left).

One might also wonder at the very large buttresses on these trees: What is that all about? Most obvious is that they are a sign of oxygen stress. The swelled buttress appears very much like the normal buttress of cypress, but they do not have normal roots. The tree's roots are at the bottom of the lake. What is beneath the buttress? Only small hair-like rootlets, a sign of stress and desperation. The buttress spreads out at the very surface of the lake at normal summer pool. The reason? The top few inches of the lake have the highest level of free oxygen, the thing the tree needs most.

Another purpose for the umbrella buttress: good fishing! The buttress is a shelter for minnows and shade; where there are minnows and shade, there is often game fish. At the right time, fishing up against these trees makes for exciting fishing, especially for bluegill.

Chapter 3

The Earliest Pioneers

Native Americans

Now the story stretches to include people, a river, and wetlands, all more unique and magnificent than any I know. Native Americans? Yes, they were here first. For thousands of years before the seismic activity of 1811–1812, Indians lived in villages and camps on the banks of the Mississippi River and Reelfoot Creek. Part of these old river bank areas are now covered by the lake, but their shores contain rich deposits of archeological information in the form of artifacts, human burials, mounds, and the remains of various animals and plants consumed by the tribes that lived here. Precisely who owned these artifacts can be difficult to answer. Several tribes were known to be in the area, including the Choctaw and Chickasaw, which might be tribes of the Cherokee. Archeologists are reluctant to say who or when the earthen mounds around the lake were built, although the period is often referred to as the Woodland period (800 BC to 800 AD). Some have dated the mounds more recent.

However, there is a clue to where might the mound builder at Reelfoot originated. Native American mound builders lived upstream of Reelfoot as recent as 1050 AD.[1] A Native American city called Cahokia in Illinois, across the river from St. Louis, had upwards of fifteen-thousand inhabitants and occupied four-thousand acres. More than one-hundred earthen mounds are found here; the main mound was ten stories high.[2] Consider this: they lived only a day-trip in a canoe downstream on the Mississippi to Reelfoot Lake.

Whoever they were, we know they spent considerable time here. Their earthen mounds are rather common. Sixty-seven sites were reported in 1986 but are now thought to number more than one-hundred, most in several locations around the lake: north of the Airpark Campground; the area south of Gray's Camp, the area known as the State Woods, Caney and

Choctaw Islands; Grassy Island (especially near Walnut Log); and along the Chickasaw Bluff east of the lake.

   The Reelfoot region was practically void of European settlers for more than a decade after the earthquakes, although French, Spaniards, and other seagoing people navigated the Mississippi centuries earlier. West Tennessee was mainly Chickasaw land. The Treaty of Hopewell (1794) prohibited white settlements on "Indian Land," a declaration confirmed by President George Washington. So the Chickasaw Indians had the Reelfoot region to themselves until they ceded it to the US Government in 1818. For $300,000, they relinquished all of their land between the Tennessee and the Mississippi River, although hunting parties were still allowed to hunt the wilderness land of Northwest Tennessee.

   Because there were no levees to hold back floodwater, the floodplain probably flooded more than once during the year, and the mounds would have been important campsites during these periods. What happened to this tribe? Tennessee archeologists report that the mound builders disappeared around 1350 AD—well before the Europeans arrived. Since early Indian tribes had no written language, their history is full of gaps. There is a clue, though. According to archeologists, a great drought occurred in the region between 1349 AD and 1350 AD, greatly affecting crops of corn, squash, tobacco, pumpkins, and others produced in the fertile floodplains of the Mississippi. So, the mound builder moved out of the region to places unknown.

   Other Native Americans were here, however, after the mound builders disappeared. The Cherokee were mainly east of here, but the Chickasaws left their name in lore and on places like the Chickasaw Bluff for the records of the first European settlers in the region.

   David Crockett was no stranger to the Indian presence. He moved from Middle Tennessee to nearby Obion River, upstream from the Mississippi some twenty miles from Reelfoot Lake. The few neighbors he mentions were seven to twenty miles apart. "It was a complete wilderness and full of Indians, who were hunting," Crockett reported.[3] Colonel Crockett probably encountered the Choctaw, since their nation was a little south in Mississippi. The Choctaw, however, were often considered a branch of the Chickasaws.

   The first white men known to enter the Reelfoot region were North Carolina surveyors Henry Rutherford with two assistants, Alex and Amos Moore, and two Choctaw Indian guides, Almer and W. Bush. Both sets are believed to be brothers. Rutherford's crew had come from Lauderdale

County, Tennessee, where they established Key Corner, a survey post set from which to reference property boundaries in the Western District of the state. James Carleton was another surveyor said to have made a survey at Reelfoot as well.

Journals and editorials by Lake County historian R. C. Donaldson are probably the most revealing records of local history and life in the Reelfoot Lake region. In a 1947 editorial, he explains that the earliest records of white men entering the area were of surveyors Henry Rutherford and his crew of four in 1785—the year they located at least eight North Carolina Grant points along the Reelfoot River.[4] Notably, the region was still part of the Chickasaw nation. The historical entry was thirty-two years before the New Madrid Earthquake. There were no roads, no lodges, no stores, and no human facilities of any kind until sometime well after the 1820s. And Tennessee would not become a state until the signing of the Jackson Purchase with the Chickasaw Indians in 1818.

Imagine the kind of wilderness these hearty surveyors encountered. At some reference point, they began their survey. Not an easy task. They fought cane thickets, cat briars, and sloughs, and zig-zagged their lines around giant hardwoods that blocked their line of sight. A baseline established later by James Carleton, referred to as the "IC Line," is said to be Church Street in Tiptonville today. All of this effort to lay out a few survey lines on property the government did not own.

How did the surveyors arrive in this roadless country? Animal trails used by Native Americans were probably the most common "highways," but the easiest way for Rutherford to reach the region was to float down Reel Foot River (from somewhere near the future settlement of Union City). The stream has been so disrupted by nature and land use that today it is known as Reelfoot Creek, although it and the Bayou du Chein (a river during this time) were very likely the highways for some of the first settlers at Walnut Log. It would be another three decades before the first settlers were known to be in the Reelfoot region.

After Henry Rutherford's entry, determining the timing of the earliest European settlers at Reelfoot Lake is difficult. But soon steamboats arrived. Steamboats brought the "modern" age of travel to the area, although I have found no record that any of their passengers hiked the short few miles to Reelfoot Lake. The *New Orleans* is known as the first steamboat to ply the Mississippi—and that was during the eventful year of 1811. Five years later, steamboats were a common mode of travel. In another five years or so organized communities were established at Reelfoot

Lake. Henry Rutherford's notes were the first to mention the name *Reel Foot River*.[5] Not until six years after the New Madrid earthquake was the western region of the state purchased from the Chickasaws (*The Jackson Purchase of 1818*). So records of human activities here prior to Rutherford's do not exist.

Colonel David Crockett and Other Early Pioneers

The first pioneers to settle in the region of Reelfoot were in awe—the country literally teemed with wildlife and natural beauty. Before them lay a new creation—a huge lake with unsurpassed botanical beauty, flush with fish and wildlife. A remnant forest of green cypress trees still covered most of the open lake; fields of yellow lotus blooms glowed in the low sun; and clear, naturally filtered freshwater sparkled for more than a mile in the moonlight. Their first thoughts were that they had found another Garden of Eden.

David Crockett was one of the very first Europeans to gaze upon the lake. No doubt by the time the colonel first saw it, half of the trees still standing mid-lake were dead snags, so thick one could hardly see through them; all of the living trees but cypress had drowned only a year or two after the lake flooded. Ten years later, most of the dead trees had broken off and would have appeared now as snags. But many of the snags broke off to become "stumps," the tops often only a few inches under water at normal pool. Yet, many groves of living cypress trees still stood, which only added to the aesthetics and magic of the lake, more than most could describe. Some concluded that only reputable poets could justly describe the inherent natural aesthetics of the lake. So spontaneously was the lake created, it seemed that all surrounding natural beauty was captured in a single moment and a single location.

David Crockett walked here with his bear dogs and camped during the 1830s only because it was one of his favorite hunting grounds. A decade later, a handful of woodsmen had stumbled upon the lake, and few left. Reelfoot Lake apparently became a waypoint for travelers headed to westerly adventures. According to the journals of R. C. Donaldson, from 1858 to 1873 or 1874, it was common to see caravans of ten or twenty wagons on their way to the West via Reelfoot Lake and the Arkansas Pole Road.[6]

After arriving, some never left Reelfoot Lake; those that left kept coming back. The family of Charlie and Lucy McQueen were such travelers. As their story has long been told, the McQueens came along from Indiana with a brood of seven children. They were headed to Arkansas when their wagon broke down before reaching the Mississippi River—a place within sight of Reelfoot Lake. Like many others headed westerly, they were likely advised of the shortcut to Arkansas from the Nashville area and east. Rather than go to the budding city of Memphis after crossing the Tennessee River at Trotter's Ferry, travelers often continued due west by way of Reelfoot Lake. They had heard of the popular Pole Road (a corduroy road of logs that lay across perennial or seasonal wet ground) and the infamous swampy land in Arkansas. But the Pole Road would help them take the shortcut through these wetlands. So, after taking Nall's Steam Ferry in Lake County across the Mississippi, the way would take them along the Pole Road, which went from Weaverville to Clarktown. From there, they could continue their journey on high ground.

But that did not happen for the McQueens. They never made it to the Pole Road, let alone to their Arkansas destination. With nothing but two mules, a meager supply of money, and little food when they arrived in sight of Reelfoot, they went to the first smoke stack they could find. It happened to be a friendly family on the west side of the lake, about where Gray's Camp is now located. Seven generations later, most had stayed on the shores of Reelfoot Lake, no farther than Gray's Camp and Walnut Log. Today, I have a string of cousins and dear friends—most still in the same area—from those generations.

David Crockett was a restless man. I think the quest to hunt black bear had a lot to do with it. So he moved from middle Tennessee to country known for an abundance of black bear, where he built more than one cabin on the Obion River. His main cabin, built in 1822, was located at the junction of the Rutherford Fork and South Fork of the Obion River. A few years later, another was built several miles downstream of the first. No doubt one of the prime reasons was to be well within a day's hike of Reelfoot Lake.

Sulfur fumes from the earthquake had hardly settled before Crockett began exploring the earthquake country. Bears were on his mind. Regarding a trip in the fall of 1825, he writes in his 1834 autobiography, *A Narrative of the Life of David Crockett of Tennessee,* "So I went down to the lake (from his first cabin, and probably to Round Lake between Lane and Sharpe's Ferry on the Obion River) which was about 25 miles . . . to do some boat building . . . until the bears got fat." When he had done that and collected meat enough for winter, he went back home. No sooner had he arrived, he writes, "when a neighbor who had settled down on a lake about 25 mile from me, told me he wanted to go down and kill some bears about in his parts." Within two weeks, they had killed fifteen bears and plenty of whitetail deer.[7] (Note: Game was very abundant back then; but by the 1880s, black bears, panthers, elk, whitetail deer, and wild turkeys were extirpated or very scarce.)

Colonel Crockett did not mention his neighbor's name. The lake he refers to on the Obion River and Reelfoot Lake are about the same distance from his home place. Later, he mentions an old friend named Davidson, who accompanied him on a bear-hunting trip "between the Obion (River) lake and *Red-foot (Reelfoot) Lake*." What is interesting about this account is that he talks about two different lakes—the only two lakes about the same distance from his home, leaving little doubt, in my mind, that at least one household was at or very near Reelfoot Lake in 1825.[8]

In her book *Davy Crockett* (1934), Constance Rourke writes that Crockett "skirted the great cracks still left by the earthquake, crossed low streams, and found Reelfoot Lake aglow with yellow lights of the great lilies [American lotus]."[9] It would have been no great surprise had Crockett built a cabin on Reelfoot Lake, but the Crockett family of history mentioned living at the lake were probably kinfolk of Crockett's son, David. Lack of convenience and time were probably the reasons he did not, because access to the lake with a supply boat would not have been an easy task had he fought the ragged swamps from the Obion River, or portaged over dry ground from the Mississippi.

It was indeed bear country. The entire region between the lake and the Obion River was perfect for black bears. It was wild and wooly, diverse in all aspects: cane breaks, swamps, sloughs, and everywhere dominated by virgin bottomland forests, many with large hollow cavities (rarely found today), well suited for bear dens. David Crockett regularly killed black bears around the lake. A couple of hunts he describes suggest that he killed at least two bears very near the east shores of the lake—with a butcher knife. One was a wounded bear and both were at bay by his hunting dogs in "cracks left by the earthquake"[10]. The wounded one, I believe, was around Bogus Hollow, about a mile east of the lake on the Chickasaw Bluff. That year he claimed to have killed 105 bears. The same year, he records an experience in which he and a friend got caught by a tremor slap in the middle of "The Land of the Shakes." He explains: "We had prepared for resting that night, and I can assure the reader I was in need of it. We had *laid* down by our fire, and about ten o'clock there came a most terrible earthquake, which shook the earth so, that we were rocked about like we had been in a cradle. We were very much alarmed; for though we were accustomed to feel earthquakes, we were now right in the region which had been torn to pieces by them in 1812, and we thought it might take a notion and swallow us up, like the big fish did Jonah"[11].

Colonel Crockett simply mispronounced the name when he called the lake "Red Foot Lake." Surveyor Henry Rutherford was there before Crockett, and he first described today's Reelfoot Creek as Reelfoot River. Subtle reference is made that a crippled Indian chief, known by Rutherford, actually camped along the river; he was known as *Chief Reel Foot*.[12] David Crockett is well known for his tall tales, and it is possible that he concocted the legend of Chief Reel Foot and the Indian maid.

Chapter 4

The Settling of Reelfoot Lake

The First Settlements and Commercial Business

By the late 1800s, there were clearly three growing communities around the lakeshore: Walnut Log, Samburg, and Gray's Camp. Walnut Log: the first recognized community at the lake; the community of my roots. Although I know of no record of the first families to settle, my family had two generations living here before my father, who was born here in 1917. Reelfoot Lake lies in two counties (Lake and Obion) and two states (Tennessee and Kentucky). It is no surprise that the lake would eventually become a melting pot of divided interests—the subject of a later chapter. People lived around the entire lake by the early 1900s.

We know that the community of Wheeling (eventually the lakeside town of Samburg) had its beginning during the early 1850s. At Walnut Log, J. C. Burdick, Jr. had a commercial fish dock in 1870 but eventually moved the business to Samburg. In fact, Burdick's fish dock was central to the future development of Walnut Log. Burdick was a mover and shaker. He had a canal constructed that connected the Bayou du Chein to Upper Blue Basin (Walnut Log Canal) and is believed to have given the community the name "Walnut Log." The canal was reworked during the 1940s. I still remember the fresh earthwork on the banks of the new canal. We do not know yet what a walnut log had to do with the name of the community. When I was a kid, I heard that a huge walnut log was used as a foot bridge to get across a wet area to Walnut Log Hotel. Could be. My grandfather said he helped haul out many board feet of walnut logs right by our front door, and many were abandoned to sink in the bayous and swamps of Reelfoot.

Samburg quickly outgrew Walnut Log. It was a matter of access and transportation. Early overland access to the lake was very primitive, but thoroughfares quickly improved around the early-1800s between the

Figure 10. A cabin near Samburg (1800s)
tolerates flooding by floating on logs.

small towns of Union City and Troy. By the mid-1800s, roads of some order
finally made it to Samburg and a grand view of the open lake. The Reelfoot
River and Bayou du Chein, along with a few dirt trails, provided the main
access to Walnut Log and the lake during these early days. Lacking roads
west of the lake into Kentucky Bend, the first farmers and residents had
the Mississippi River, which bustled with trade. But the few miles between
the river and Reelfoot were still an uncertain wilderness in which cow
paths were probably the best roads.

New opportunities were on the way. By the later part of the 1800s,
tents and rough shelters were still around but fishing hamlets like Walnut Log, Samburg, Kirby Pocket, White's Landing, Champey Pocket, and
Gray's Camp had begun to be more permanent. Business buildings were

upgraded, one house became two rooms, fishing piers were built, and some buildings were anchored on rafts of logs to rise and fall with the floodwater. Sure, that was progress. It might seem rather convenient to tie up one's boat and fish right off your front porches. Utilities, of course, were still no better than wilderness camps—kerosene lamps, wood stoves, water buckets, one-person wash tubs for bathing, and airy outhouses were convenience enough. Bathing? It was tricky during winter; trying to sit in the middle of a metal number 8 washtub and not touch the ice cold sides of the tub was unpleasantly cold, even during summer. The consolation? Well, half of the nation was not much better off.

But Reelfoot Lake was truly an island of plenty, if you didn't mind a bit of primitive living. The lake was a diverse wetland, replete with fish and wildlife and a vast oak-hickory forest around it. Mushrooms and wild berries were plentiful, and a wilderness of swamps, canebrakes, and rivers were full of natural resources—everything a woodsman needed for a happy life. In his 1834 bibliography, Crockett called it "Red Foot Lake"[2] Crockett was one of the first to report the abundance of wildlife in the Reelfoot Region

R. E. Lee in his *Reelfoot Lake* book of 1915 states that a Mr. Campbell (father of George C. Campbell of Campbell's Point) "came to this section of the country from the State of Illinois in 1841, and boarded with a gentleman by the name of John Mitchell. He hunted and fished for a livelihood, and lived here since boyhood, long before the Indians had abandoned this section of the country."[3]

But the Civil War significantly changed life around Reelfoot Lake. The war damage was one thing, but during the recovery, outsiders became astutely aware of its bountiful natural resources. Most came to take advantage of the lumber industry and great hunting and fishing opportunities. Perhaps more than two-hundred families lived at the lake by the 1880s. Only a few miles southwest, youthful Tiptonville was recovering as a river port, thanks to the migration of the Mississippi River channel and the Union bombardment during its victory at the Battle of Island #10.

You cannot hide from war; war even affects the wilderness woodsmen. The Civil War battle over Island Number 10 was barely more than two miles from Reelfoot Lake. It was furious and short, lasting from February 28 to April 8, 1862, but the massive troop movement in the region disrupted everything in its path. So, it took time to readjust their lives. They did adapt, but the simple life began to erode at its edges as the modern period approached.

The good life far out-weighed some social changes and troubled times. Yet trouble always seemed to lurk. What I consider the best years at Reelfoot did not arrive until after the early growing pangs of the 1900s, when competition, politics, and legal matters entered the quiet communities around the lake.

Who owned the lake was a nagging, inevitable question. Those who had the right to use it were landowners whose claims were justified by courts. But the landlords also claimed exclusive rights to the use and income that accrued from the lake. Most Lakers did not agree. They had settled this land, raised families, and built a society of peace and good will, and now this: their rights to a livelihood had been abused, and they took action. In the heat of desperation, a knot of men formed a vigilante—the Night Riders, who played a key role in the ownership and future of Reelfoot Lake, albeit with questionable justice for some and a deadly outcome for many.

When the lake was declared public domain by the State of Tennessee, the best years at Reelfoot Lake began. By the 1920s, memories had faded and not much was said about those troubled times. The Lakers, often referred by now as "The Gumbooters," enjoyed some wonderful years of prosperity. The 1940s and 1950s were the years of my youth at the lake. Times more peaceable began to change again sometime after the 1950s. It lasted into the 1980s when concerns about the management of the lake rose to a minor crisis. New wildlife laws restricted the taking of fish and game; social mores were transitioning to a more fast-paced world; changing markets demanded new skills—often different from the old—and attempts to make significant changes in the management of the lake were rejected by locals. The transition changed a century way of life for Gumbooters, considerably.

Still, I can imagine no better life for a kid than to grow up with nature as I did in my youth. Reelfoot, to me, was the epitome of all the grandest creations God ever made. I knew nothing about conflicts; conflicts of yesteryears were just another old story. I could hardly express myself better than the earlier words of Mr. Lexie Leonard: "When I'm not on the lake, I want to be," he said, "I have to be very careful not to worship the creation more than the Creator."[4]

Chapter 5

The Gumbooters

What else could they be but hearty pioneers who loved the freedom of living amid the wilderness. Although time encroached upon this wild, wild land with more people and commerce, the heart of Lake People changed very little for over a hundred years. These were the Gumbooters. The Gumbooter's passion for the great outdoors and way of making a living went right along with being free to choose his methods, and he used them at his own pace. it was at the roots of his being from the day he was born, and he passed it on to his wife and children. Instilled with most of the virtues and faults of the first pioneers, Gumbooters would be opening Tennessee's last patch of wild frontier.

The First Generation

They were few and widely dispersed. As wild as the country could be, the notion of civilization had preceded David Crockett by a mere few years— by the 1820s, upgraded roads were already being planned to tie Hickman, Union City, and Troy to the cities of the east. Crockett, however, had gazed upon the waters of Reelfoot Lake before the first fledging settlements of the 1850s. One of his biographers, Constance Rourke, wrote that Crockett "skirted the great cracks left by the earthquake, crossed the low streams, and found the waters of Reelfoot Lake aglow with the yellow lights of the great lilies."[1] You can still see them today. The beautiful yellow lotus blooms in mid-summer at sunrise and sunset fit the description quite well.

Crockett, on his black bear–hunting expeditions, encountered only a few Native Americans in hunting parties and a pioneer or two living here. In 1835, most of West Tennessee was still a frontier, in fact, largely undeveloped, although clearing every patch of wilderness possible soon begun in earnest. "Clearing the Back Forty" was still the thing to do when I was a kid, although I might have been the last kid to hear that phrase.

By this time, David Crockett had explored Northwest Tennessee and had released his autobiography; Charles Darwin had reached the shores of the Galapagos Islands; Mark Twain was about to be born. It would be another twenty years or so before Captain Mark Twain would pilot his steamboat around Kentucky Bend, so close to Reelfoot Lake he could almost see the ancient, giant cypress trees that grew here. Aside from a few landings and family farms in Kentucky (Madrid or Bessy) Bend, the region of Reelfoot Lake was still as much a wilderness as it ever was.

Rourke writes that once at the head of the lake (probably Upper Blue Basin in the 1830s), Crockett and his son passed a little thatched hut, where they caught a glimpse of an old hermit living there. "We want bother him," Crockett said, "Since he wants to be alone. I know that feeling well."[2] It was Chickasaw land, and remained so until the Jackson Purchase of 1818. While there were no known villages in West Tennessee, the Chickasaws continued to hunt here until the Trail of Tears in 1837 forced them to leave for Oklahoma.

"The Land of Shakes," "Woods Lake," "The Swampland"—all were descriptions of Reelfoot Lake. The region was wetlands and forests, very remote until the communities of Walnut Log and Samburg appeared on maps around the mid-1800s. Sometime around the 1840s, the first settlements appeared at the lake by the recognition of J. C. Burdick's commercial fishing interest at Walnut Log and the Wheeling community or Samburg. The majority opinion of the general public, however, was that Reelfoot Lake was still too rough to live in. The glowing reports of natural beauty and the abundance of fish and wildlife had barely preceded the Civil War; afterwards, Reelfoot Lake was soon known as nothing less than a providential hunting and fishing paradise.

Lakers were the Mountain Men of the Reelfoot region. Most Lakers gave little thought to commercial enterprises other than commercial fish and fur markets. Living in a remote wilderness was their joy and livelihood. Like their counterparts, the Mountain Men, they were likely content wanderers, hermits, or eternal woodsmen, many with no families and no mission in mind, other than to live isolated from their past and from encroaching civilization. That luxury soon vanished. Sportsmen and adventurers with better means discovered the great wealth of hunting and fishing resources at the lake.

Colonel David Crockett was ecstatic that the northwestern part of the state was still a wilderness when he arrived during the 1820s and hunted through the 1830s, still with copious populations of black bear and other game, a place to escape the cluttered life of politics and farming for a living. He took every opportunity to hunt the low lands of West Tennessee. Crockett could have been the mascot for these early pioneers. He was a self-made naturalist. He realized that the lush, bountiful hunting grounds he so loved existed because the native forest had an enormously diverse plant habitat. Overharvest, however, did not seem to enter his mind when it came to hunting bears. He should have: the effects of over hunting and "progress" east of him were the reason he moved to the wilderness of northwest Tennessee.

Reelfoot settlements were not the first in this remote region. Small communities and a few family farms immediately west of the lake in the Kentucky Bend area preceded settlements at Reelfoot Lake by a few years. These were generally farm families and merchants with facilities on the highest ground along the eastern banks of the Mississippi where a growing steamboat transportation industry was in the making. The threat or reality of some kind of disaster always seems near when living in river floodplains—but the stigma of it is soon forgotten and life goes on.

The temporary and fledging river town of Tiptonville a few miles downstream of the Bend saw its first buildings in 1857, some of which were soon moved or fell into the land-hungry river. It was hardly recognized as a town before collateral damage from the Battle of Island Number 10 in 1862 resulted in its ruin. Rebuilding began that same year, only to suffer the Great Fire of March 19, 1901. Fanned by a strong southwest wind, the fire practically destroyed the central part of the town. The river claimed the newly established town in 1905, after which it moved a short distance east to its present location. Today, Tiptonville is a sleepy town two short miles west of the lake.

The frontier towns of Union City and Troy were at the eastern edge of the northwest wilderness. They had become recognized but fledgling communities by the 1820s, about the time Memphis was being recognized. Located some ten to fifteen miles east of the lake, there were no travel ways between the lake communities and towns better than cow or game paths, and miserable and muddy wagon trails during spring and most of the winter. It caused the Lakers to be frugal with what few supplies and dry goods they could get, much of it packed in on mules and their backs. Riverboats had long plied the Mississippi but they were not too helpful because landings were remote from Reelfoot. The nearest was Slough Landing, located in New Madrid Bend not far upstream from today's Cates' Landing. It might as well have been the Port of St. Louis insofar as the communities of Walnut Log and Wheeling were concerned, since there were no roads between them.

 Solidarity was a passing thing for the earliest settlers. The rising reputation of this paradisiacal lake changed all that. A growing lake society was on the way. Characteristically, The Lake People spent most of their lives working or recreating in the wetlands of Reelfoot Lake—the Swamp People. It was just another label outsiders sometimes used for the Lakers. Whatever they were called, they lived freelance, making a living their own way from natural resources at the lake and surrounding land. They knew no government boundaries; state or federal county lines made little difference to them, even after boundaries were established in 1870. They stood together as Lakers, somewhat skeptical of new arrivals with starched white collars and business ambitions; starched collars did not seem to fit a rugged life in which most men wore gumboots from Monday through Sunday.

 High water and swampland shaped much of the settlement activity at Reelfoot. Hardly enough folks could be found in one location to call a town until Samburg was recognized. Wheeling was the first settlement that took notice of the Mississippi River floods and built the community on high ground at the foot of Chickasaw Bluff. But that changed. The lure of the lakeshore was too great. Wheeling relocated to the lakeshore and was renamed Samburg, where it remains today. The new location allowed a panoramic view of the lake's extraordinary natural setting, and afforded immediate access to the lake, which was an enduring comfort to the citizens and better for business. Samburg became the largest settlement around the lake. Other communities settled around the lake for the same reasons. Two standards were common development requirements for a community; a knock-your-socks-view of the lake and terra firma at least four feet above lake level—"high ground."

 The "Jamestown" of Reelfoot Lake was Walnut Log; it was the first recorded settlement. Its residents seemed to ignore high water. Its early establishment was most likely because, in the absence of roads, two rivers provided ready access. The Bayou du Chein and Reelfoot River were legitimate rivers before the New Madrid earthquake. Deep and wide, these rivers could float sizeable water crafts with heavy loads.

 Next was the community of Wheeling, soon to be Samburg. Located south of Walnut Log by five or six miles, more land suitable for development was here, and eventually better roads provided access from Union City, Troy, and Wilsonville (future Hornbeak). A road from Samburg to

Tiptonville would not be forthcoming until 1922–23. The delay was caused by wetlands like "The Scatters (biologically rich wetlands immediately downstream from the lake)," a spillway outlet, Reelfoot Bayou, and other wet areas, which were hostile to road-building. That challenge was overcome with a rather primitive road as part of the 1919 "River-Reelfoot Lake-Levee-Road Project" around the south and west end of the lake. Of course, one of its purposes was to regulate the level of the lake.

Still, it would be a stretch to call Reelfoot Lake an industrial complex. Yes, a lumber industry soon began, and it did employ a good number of citizens, but it was a finite industry, and not one local bread winners could depend on. The future was based on the hunting and fishing business, a resource that seemed to have no end. The tools for this trade were sparse essentials for the average Laker: a few tools, a wooden boat, a twelve-foot push pole, a landing net, trot lines, or webbed fishing gear (which they made) were was about the normal Laker inventory. By this time, *gum boots* left more tracks along the shores of the lake than any other footwear.

Fishing was their livelihood. Indeed, fishing was so productive through the next few decades that some were inclined to use a simple cane pole, hook, and line for commercial fishing. Nylon twine (which had a long life) had yet to come on the scene; nets were knitted from cotton twine, which lasted only a season or two. Consequently, it took considerable time to knit new cotton nets and to preserve them in vats of hot tar for a longer life. Summer days were often permeated with the nostalgic odor of boiling tar vats. Even today, a whiff of hot bitumen brings back youthful memories of Reelfoot. Commercial fishermen dipped their cotton nets in these vats and stretched them to season in the open air before fall fishing began. In addition to their nets, a few dozen steel traps and three or four custom-made fish and frog gigs were enough to make the Laker and his family a living.

Hunting: was it work or sport? Probably both. Game was not only a daily source of sustenance, but also a major source of income, either through guide and lodging services, or selling wildlife products. It was not uncommon for a family provider to rise before the sun to collect game table fare for the day, including a mess of squirrels for breakfast! Waterfowl hunting was champion of the hunting business. The collection of ducks and geese for the market might have provided more income to the average Gumbooter than guide service before bag limits and other restrictions were legislated by the Migratory Bird Act of 1918.

Homes during these early years improved very little, except in the budding town of Samburg and, perhaps, the lodge at Walnut Log. Tents were not so common anymore, but shanties—for the most part made of whatever could be found—were common enough. Preferred building material for houses and boats was from the ancient heartwood of old cypress trees, usually available from local sawmills; it resisted rot and made for a very warm interior for houses, businesses, and other buildings.

But as the last decades of the 1800s came to a close, the world around them began to change. Subtly, innocently, change was disguised as progress. But regulations, transportation, and the farming industry encroached heavily upon this Garden of Eden—changes that were also taking place on a national level. But the people of Reelfoot Lake faced their own new challenge: Would they ever learn how to live at ease in a river floodplain, realize the long-term value of its rich natural resources, and not destroy the hand that feeds them?

Land ownership was a ubiquitous problem. Most of the new communities owned no land nor had any idea who held title to some of the land they lived on. The criteria for building anything was to find ground high enough near the lake to keep their woodpiles dry and boats moored at their backdoors. The Mississippi flooded nearly every year, usually during late winter or spring. Since there were no levees substantial enough to hold back floodwater before the early 1900s, when the river was flooded one could ride in a boat from the Chickasaw Bluff to high ground in Arkansas or Missouri.

The answer to flooding: levees and levee districts. Levees to "protect" the area were another problem; everyone, it seemed, wanted them. It was a prescription for levee wars. Never mind those that easily survived these floods because they built houses on piers well above floodwaters. The earliest levees were private levees constructed west of the lake to protect farmland, not houses. These were mainly in the Kentucky Bend area. There was another problem: levees were already being built on the Arkansas and Missouri side to protect farmland, obstructing the floodplain and forcing floodwater to Lake and Fulton Counties. Tennessee and Kentucky landowners obligated to step up their

Old photographs showing marks on cypress trees indicate that one year Samburg stood in about eight feet of floodwater—a lot of water. But the government thought it could harness it, tame the wild stallion called the Mississippi River, and make the water do as it pleased. Levee proposals had long since been on the minds of Tennessee farmers along the river. Building levees had become a big, incautious business.

The establishment of the Lake County District and the Fulton County, Kentucky Levee District in the early 1900s was notification that the race to build twenty-two miles of Mississippi River levees (seventeen in Kentucky and five in Tennessee) was now underway. The levees would protect some 175,000 acres of overflow land in these districts, rich land that might be farmed and support thriving towns and communities. There was little or no thought as to how these embankment might affect the natural resources that attracted the first generations of lake users in the first place. The completion of the levee in 1909, however, was not much comfort for Lakers and landowners three years later: the 1912 flood reminded them just how puny levees could be against the might of the river.

Soon after the levee wars began, the government became involved. The Tennessee Legislature passed Joint Resolution 38 in its 1859 session, directing its congressmen to get the federal government involved. A counter proposal to the Arkansas levees: build levees on the east bank of the Mississippi from Fulton County to the Wolf River at Memphis. And so began an all-out war against wetlands below Fulton County, Kentucky, some forty miles or so to the Obion River. Battles raged in many places along the industrial-minded communities of the Mighty Mississippi.

Next the war was against the "hazards" of wetlands. The new strategy: drainage districts. Following the levee districts came the formation of the 1913 Drainage District for Lake, Obion, and Dyer counties. Its purpose was to rid the counties from the meanness of "the Scatters," all wetlands below Reelfoot Lake to the Obion River. We find no mention of how valuable these wetlands were to the Reelfoot Lake, its fishery, migrant waterfowl, or anything else; to the world beyond the Lakers, these wetlands were nothing more than reservoirs for snakes, mosquitoes, malaria, and other diseases. Fortunately, someone had the vision to see the predicted fate of the Scatters.

Lake Isom was a southern extension of Reelfoot Lake. In 1938, the federal government bought 1,846 acres of the Scatters, which became Lake Isom National Wildlife Refuge. Today, it is surrounded by farmland. The forest has been skinned to raw dirt for miles, but the small refuge precariously remains as an effective waterfowl refuge—although threatened by sediment and chemical deposition from soil erosion and runoff from adjacent fields.

While farmers saw floods as disastrous, Lake People saw them as inconvenient. They knew inconvenience. They knew high water. They knew how to adapt. The woodsmen were the friends that usually bailed them out of these inconveniences. Most had the skills and could craft and build the necessary facilities, including boats and houses that floated or sat upon piers. The best tools were not always guaranteed. The backs of mules—or their own—were the most dependable source of supplies until roads were fit for the drummer's wagon. It was one reason they had few crops; crops were generally limited to a patch or two for gardens. Indeed, the Lake People had no rent, no newspapers, no electricity, no indoor toilets; they had no ice boxes, no taxes, no constables, and many were so isolated they had few neighbors. A fortunate few had candles or a kerosene lantern. Survival and security, however, did not weigh heavily on their minds; it was a way of life.

The pace of life changed rapidly after the Civil War, which "officially" ended on May 9, 1865. For some, the times changed too fast; many were pleased with what they had. But there was no turning back now. Once a few roads were built, the progressive life of the Laker moved along at a fair pace. By now, the wilderness days were but good history; it was high time to settle in and enjoy the abundance of this great Garden of Eden.

Those who daily reeked of swamp water and fresh fish began to view their circumstances as a path to a desirable and reliable future; that, of course, was only with a view that nothing would change, and that this cornucopia of nature would continue and last many lifetimes. Yes, it was a slightly shortsighted view. But they were mostly concerned at that juncture about a place to live. Finding a suitable and available location in which to settle was not so easy. First, it needed to be right on the shores of the lake where a boat might be moored within a few yards of the house. Next, it needed to have a good view of the lake, which kept the proprietor's soul calm and his mind on what to do the next week or two to feed his family. He could do with less, but his heart was set on doing a little better. Planning further than that, it seemed, usually had to wait its turn. The problem was that sloughs, marsh, and cypress swamps were more abundant than high ground. Nevertheless, anyone could see that Reelfoot Lake was still dependable and teemed with fish and wildlife. So the lake community settled in, and eventually saw even better days.

Was the lake made for these pioneers or were they made for the lake? Hard to say. For the time being, there was certainly enough for everyone. But it would not last. Such rich natural resources and way of life could not be hidden from the world for long. Obion County had the state's blessing to build a few passable roads in the interest of commerce during the 1820s and 30s. Before long, local citizens would be afforded the pleasure of not only better roads but also more commerce. Reelfoot Lake could be a tourist destination, a place for hunting, fishing, and other outdoor recreation activities. And that's what happened.

Another generation passed. By the 1940s, the floods of yesteryear had faded into dim memories—for the young, but not so much for the old. Tents, shanties, and buildings on floating logs were pretty much relics by now. Substantial cabins, lodges, and houses made of cypress board and batten, and the foundations of some on elevated piers, remained evidence that the floods of 1912, 1927, and 1937 were not to be ignored. The Tiptonville Dome might have been the only ground between the bluff and the Mississippi not covered by water during those floods.

So what? Lakers adapt. Some built their houses on piers to be above flood zones, some seemed to have coercive amnesia. Others move to the foothills of the Chickasaw Bluff. But most seem to have put up with high water, many of whom moved the edge of lake into conventional houses with the rest of society. Some became farmers during the summer and hunters and fishermen the rest of the year. The towns and communities of Samburg, Walnut Log, Blue Bank, Champey Pocket, and Gray's Camp were historical settlements by this time. Tourist and sports hunting and fishing interests created very active business communities along the lakeshore. Yes, it was a colorful life and a wonderful era at Reelfoot Lake— and I was there to witness much of it.

Gum Boots and Gumbooters

A period of transition arrived around the second decade of the 1900s; it had a lot to do with clothing. One outstanding advance was the availability of a good pair of gum boots. Extremely practical gear and superior to leather boots, the footwear was usually worn with the tops rolled down to the knees in a "Puss-in-Boots" fashion, and pulled up to full thigh-length when needed.

The name was appropriate enough because they were made of real rubber, gum from the Amazon jungle, natural latex, sap from rubber trees, distilled like good Kentucky whiskey. Not so today; hip boots hardly last a year. The quality of gum boot footwear back then was immensely superior; they did not readily deteriorate from ozone-rot, and punctures were easily patched. Most of these boots, patches and all, were passed to sons and daughters. It was perfect day-to-day footwear for most for these reasons; a perfect boot and perfectly matched to the character of its owners. I had one objection—the boots I had were not insulated. My feet, unlike

Dad's, stayed numb from winter cold. I often wondered if my toes would be permanently blue, if they didn't fall off. That finally changed with the newly designed, synthetic rubber, insulated boots—just in time to see the last of real gum boots.

I doubt if my great grandfather had a pair of gumboots until late in his years. Since his day, there were some upgrades besides gumboots: better shelter with fewer leaks, better wood stoves to cook on, kerosene instead of bear grease lamps, stump-jumper boats had improved; they had better fishing nets and carbide headlamps for night fishing and gigging bull frogs, more kitchen wares than a skillet and pot, a few condiments, markets for their fish and game, and, of course, gumboots.

Those were the wonderful years. I missed the earliest years of that period of lake history, all because of a simple mishap in the year I was born—one generation too late. Still, I was fortunate to have lived during the very last of those great years. But through stories of elders, I lived them all—virtually. I don't think we ever thought about any other life. Not that it mattered because matriculation in hard knocks degrees was about all we could brag about. To have a starched, stiff-white collar and necktie occupation with a mansion on a hill would have been too much like a city life. Anything of that sort would surely have been anathema to the very life we led.

There's a lot to be said for living in comforts of a weather-proof shanty. With a potbellied stove and a little screened-in porch, a hunting dog for a pal, no rent, no taxes, and not even a title to the land you live on—and no objections from anyone. What was there to need? We had the freedom of our Choctaw brothers and enough inheritance from our European fathers to live a near-optimum life. The smell of fresh fish, curing furbearer hides, and a burnt whiff of gunpowder on our coats still drifts by on the breeze of certain fall days for those of us who still live here. Just nostalgic memories, memories and stories brought back like a warm spring shower, and all of them real enough to include me. That's the way it was well into the 1950s—a gum boot life.

Although Gumbooters were put together with raw hide and tough sinew, they were mostly people of good cheer, and their life style often much envied by visitors. Young or old, the true to their breed that I knew needed little incentive to be on the lake. Even though some might have been a bit short on ambition, idling away would not last long because the lake beckoned. So even though markets were down or glutted, they might go right on with their occupation, just to defy the world's troubles and to keep a calm soul. During these times, they took pleasure in sharing

Figure 11. Carver, commercial fisherman,
guide, trapper, naturalist, and gardener
"Ras" Johnson raised six happy
Gumbooters with wife Mildred.

their catches with friends and good neighbors, something that seemed inherent to their raising. Not uncommon was a promised mess of fish to a fish-hungry friend or neighbor, even before the nets were raised. This was simply part of their nature: generous and free-spirited, and embellished by the freedom to do as they pleased; they had not a single foreman to tell them when, what, or how to do. Most considered themselves as rich in soul as any man. Rich men often concede that very point. It was surely a rather healthy way of living because most who survived ordinary calamities lived to a ripe old age.

Freeze-ups, Refrigeration, Bread-'n-Butter Skills, and Paradise

But even a good life faces times when making a living could seem a bit strained. Long freeze-ups and nowhere in mind to vacation, however, could half-deplete their larders. Freeze-ups around Christmas often did not thaw until late February or March. That could mean fishing, waterfowl hunting, guiding, and most trapping was all but a bust. Nevertheless, fish for the table was a staple as important as bread and meat. For most, the only refrigerator was a block of ice from the nearest ice house. Before that convenience, blocks of ice were cut from the lake, covered with sawdust, and buried. A chunk of lake ice might last through most of the summer. Game and fish might be stored outside during long, cold winters. I very well remember some of those long, hard winters. We hung freshly picked and dressed waterfowl outside—preserved by the chill of winter for weeks—their bills snapped to a line of nails on the back porch or shed until duck and dressing were on the menu. That seasoning made the better duck and dressing than money could buy. Waterfowl and fish were staples during winter. Freeze-up or no freeze-up, fish was always on the menu. A few nets were left in the lake to satisfy that household requirement. It meant chopping a hole in the ice to free and empty the nets. It was more fun than work.

So Dad and I would hitch ourselves to the lightest boat and haul it over the frozen lake where the barrel nets had been set. Sometimes, we would put steel runners from net hoops on the bottom of the boat. Like a skate, the boat skidded easily over the ice. The lead "pusher" had to be careful not to be run down by the boat since it could outrun us. An excellent safety backup, it could also carry a heavy load of fish. Chopping and sawing a hole in thick ice large enough to haul up the net might take an hour, but it was usually worth the effort. Just about any fish would do when you were fish-hungry, except maybe gar, drum, shad, or bowfin, depending entirely on our hankering for fish. One catch might be enough for several weeks if temperatures remained low enough outside to freeze, which preserved them. Good fishing tackle was always a reliable backup for hard winters. Besides, it took only a few days of thawing before boat trails and wind broke up the ice and fishing, hunting, and trapping went on as usual.

Guide service for hunting and fishing, commercial fishing, and trapping remained the bread-and-butter skills, even during the heydays of the lumber industry. Most made a reasonable attempt to abide by the game laws. Lawyers were expensive. Gumbooters avoided them to the extreme, even though they sometimes needed them. The rowdies, who infringed upon the law, I think, didn't fear the sheriff nearly so much as they did the lawyer fees.

Generally speaking, one's word was as good as a contract. Of course, rules hardly applied here. Gumbooters were quick to suspect men of this nature and their lot was unwelcomed and hardly worthy of mention. Not that it was unheard of to have a killing, but the core of hardworking Gumbooter avoided the ruckus of that breed. Honesty (even the rare thief did not deny his occupation) and fair dealing was their hallmark, which did not count when strangers took them for granted. Yet, one thief I knew stole from his neighbor's garden. I asked him why he took advantage of the old man and his wife, since they depended heavily on the garden for groceries. The thief's logic was that they should not have put the garden where he could see it. There was justice in leaving a few of this breed stranded on a stump in the darkest swamp.

Still, Lakers greatly enjoyed conversing with strangers. Those who visited the lake, even a perfect stranger, might well be invited to a Laker's table before the day was over. From then on, they might be as welcome as a first cousin. Many "sports" became life-long friends and part of the family.

Lack of fish gear for the commercial fishermen could rarely be an excuse not to be fishing. For those who too often mishandled their gear and income, fishing gear with obligations was available from the local "Godfather," who controlled the buying and selling within the commercial fish business—and the fisherman's credit. There was a cast of men who stayed indebted to the Godfather their entire lives, which naturally was intended, for these fishermen were the main source of his political influence.

Every season and holiday was to be celebrated in some fashion. Most ambitious Gubooters trapped for the fur market and forged the woods for herbs and mushrooms when other activities were at a lull. Once a month or two, we went to Tiptonville or Hickman, Kentucky, to sell furs, fish, or ginseng and yellow puccoon roots. There, we'd do the essential shopping; basics like ammo, traps, net twine, tools, and condiments for the larders. Communities were small; although separated by five or ten miles, families knew each other, or at least they knew each other's names. Community celebrations were more often a Fourth of July BBQ. Preparing for the BBQ was a community ritual that began at least two days before the holiday with the dressing of a fat goat or half-grown pig or two. A large hickory-kindled bonfire was set sometime before sundown. By dark, a glowing pit of coals signaled time to begin the roast. The dressed meat was tended off and by the men folks every hour day and night from that time until about noon on the fourth, while the community children gathered for games and others looked on until the wee hours of the night. With a few guitars, a pair of clacking spoons, song, and dance, all had a jolly good time. It was probably the most enjoyable celebration of the entire year—and the best barbeque north of Memphis.

Families enjoyed picnics at the CCC's Round House or Washout beach, or neighborhood fish fries. Permanent buildings for churches were limited, but that did not deter believers; old-time camp meetings under a brush arbor were put together with little notice. As David Crockett might say, "They could slap up a church in no time." It was a life in which fishing and hunting gear could sit exposed in a boat for a week and not be disturbed. A legitimate need for help was always answered.

So it was, for a hundred years: Reelfoot was truly an outdoor paradise, teeming with fish, wildlife, and natural resources, all that a natural oasis could be. It was a period when Gumbooters with a little get-up-and-go could make a decent living for their family any season of the year.

Winds of Change

Changing times were apparent by the late 1950s. Soldiers from the war had come home less than a decade earlier not sure if they should pick up life were they left it, or take advantage of GI bills for a new life. By the end of the decade, families had grown, industry was on a roll, and Gumbooter fathers began to find that their former occupations had constraints not expected—markets they depended on for the lake's declining resources could not readily support new trends and needs. Mothers were more ready to adapt to modern life, and the young'uns of that generation were prepared for the change. Schools had improved curricula and began to consolidate at locations farther from the lake. The students were not all Lakers and the mix consolidated new ideas and life styles. Still, the affectionate names of "Gumbooters" or "Lakers" were accepted locally without offense, for lo, these many years. Lake people who lived a little too far from the shores of the lake, including the Chickasaw Hills and communities therein, were a reach too far to be genuine Gumbooters—these were "Kneebooters," a little less esteemed than a Laker, and a little less tolerant of their given title.

The time of isolated homesteads had seen its best days. Community life had become the preferred norm; neighbors were family, and if fate forced the family to move even a quarter of a mile from the community, life could become dreadfully lonely for mothers and children. As for my family, if there was a gene for being a hermit, I believe it had finally been lost.

I remember that feeling quite well the year Dad decided we should move a few miles from the lake and try farming. Socked into a backwoods farm, I could no longer see lake water. Like a wing-clipped duck in a chicken cage, it took a slice of happiness from my life. Without immediate access to the lake; without the nearness of good neighbors and Huckleberry Finn friends, something as essential as freedom had been sacrificed. At this point, we were now truly as pioneering as the earliest generation of our forefathers.

Walnut Log and the Golden Age

Walnut Log was a minor community that might have characterized the Gumbooters of Reelfoot Lake. During the late 1800s, it had a dozen or so houses and peaked at a little more than twice that during early decades of the 1900s. The community was blessed with a two-room grocery store for as long as I can remember—one always owned and named by the multi-generations of McQueens. The first grocery (early 1900s) was probably owned by Alfred Lyon, whose daughter was Lillian Crossley. Adding to the color of the community were boardwalks to cross the Bayou du Chein, a commercial fishing dock, and Ward's Lodge, followed by O. T. Wallaston's Lodge, very active sportsmen's lodges and often central gathering places for resident activities. Walnut Log, as you might recall, appeared in local history a short time before Wheeling, which became Samburg. (A village of perhaps one-hundred residents during the early 1900s, it had not exceeded 210 residents by 2010.) These were the only recognized communities along the lakeshore during this earlier time, although there were individual enterprises on islands of the lake and other isolated places like Lake Center, Della's Towhead, Caney Island, Allen Basin, probably Gray's Camp, and others that remain a secret to this day.

Unlike most lake communities eventually established around the lake, Walnut Log did not have the magic that comes with a view of the lake. No. But it was more than a little unique—it had the Bayou du Chein, which was once a viable river. Bayou living was not so unique in Louisiana, but it was in West Tennessee. The river once drained a good part of Western Kentucky and prowled through Reelfoot Lake and on to the Mississippi before the lake was born. Sometime around 1800, as a best guess, the Mississippi River muscled its way a bit too much east and literally stole a segment of the Bayou du Chein River. That segment was a good stretch immediately west of Hickman, Kentucky. You can see the upper half of the river by the same name today where it empties into the Mississippi River not far above the town of Hickman. Of course, that left not more than half of the river remaining to go through Tennessee. Most of its energy was lost by the time, the Bayou du Chein River went through Walnut Log, and so did its title as a river. I suppose that was reason enough for most locals to call it "The Slough." But the crippled segment was still fifty to sixty feet wide and eight feet deep when I was a kid. Because its watershed has been drained, ditched, and diverted, it became something of a quiet stream, which was not always an accurate description because it could still get rowdy during heavy rains. So, the river became simply the Bayou du Chein—or "The Slough" to locals—although it was not technically a slough. Fortunately, it left banks high enough through the community to build houses and keep a dry foot path—unless the Mississippi got on a rampage.

Walnut Log was quite a unique community in those days, sleepy some days and lively on others. Not to be excluded entirely, Walnut Log Lodge enjoyed a view of the lake; from here, a little vista of the lake can still be seen at the other end of the Walnut Log Canal—Upper Blue Basin. Of course, the lodge is gone today but a concrete boat ramp is still at the lodge landing. Standing at the boat ramp (the site of the old lodge is directly behind you), you can still have about the same view as the old timers after the original channel was re-dredged sometime during the 1940s. Look to the right on the opposite shore of the bayou, and you will notice tall cypress trees; they were a feature of the community swimming hole. A 4" x 10' x 16' oak diving board was attached to the largest tree. You could say we were deprived children, but that isn't necessarily so. We didn't have ball fields large enough for anything but sock ball, no skating rinks, soda shops, or tennis courts—but we had plenty of nature to entertain us. Just about every Sunday the young (and young at heart) would gather here for a swim and an outing. McQueen's Grocery was fifty yards or so up the bayou. So, an RC Cola and a Moon Pie, or an Orange Crush with a sack of Planter's Peanuts were favorite refreshments after a short swim or casual sunbath. Back then, the bayou was about twenty feet deep at the swimming hole. Cypress knees along the bank were large enough to make benches. It was a spirited little community every Fourth of July with laughter, games, kids romping, and the fragrance of pit BBQ and water lilies drifting down the bayou and permeating the woods and marshes for a mile.

Walnut Log had twenty or so resident buildings along one side or the other of the bayou with individual boat sheds, fish, and turtle live boxes to accommodate each. Another half-dozen resident homes were along the dead end, dusty gravel entrance road to the bayou community; most were hardly more than shouting distance from the bayou. Walnut Log Road today is a shortcut across a former loop road. The road once took a longer route up the bayou a quarter mile (B. Johnson Road), turned east to the bluff, and looped back south to the Reelfoot-Hickman Road (Highway 157) where the existing road begins. Land along the road during those years was open land. It was purchased by TWRA during the 1980s as part of a buffer zone around the lake. Today, it has been returned to forest land.

Out our back doors looking west was a strip of woods and then the marsh, fifteen-thousand acres of open lake, and a former river running through the middle of it. You could say we lived *in*, rather than *at* Reelfoot Lake. The bayou and canal provided quick access to the lake, and protection against storms and high waves that could wash out our docks. Yes, although it might be difficult to imagine, this was once a bustling community with a grand lodge painted green with white trim, a country store or two, and often a seasonal cadre of scientists in and out of a mysterious building called the "Bug House."

That was still the "Golden Age," a period when making a comfortable living was as easy as "falling off a slippery wet log." It was a magic place with so many things to do you could not get around to doing them. Game hunting, testing a hot fishing tip, a summer ginseng romp, a trap line to run, meeting anxious sports at the lodge, and so on, were part of daily life. A fever of anticipation and enthusiasm always permeated the air around us. For me, breakfast was gulped down as if I had to catch a train ready to pull out of the station. The entire community seemed to be upbeat most of the time. It lasted until I was a teenager; as I've said, until about the mid-1950s. Who could want more?

Reelfoot Lake had long since been a destination resort where lodging and guides were top services. By the 1920s, Walnut Log was a prime destination for sportsmen and outdoor enthusiasts. Samburg (Wheeling at the time) had risen to a lively community at the foot of the bluff a decade or two later. Not long afterwards, it moved to its present location. Dirt streets turned into graveled streets, and the hamlet grew. Stationed right on the bank of the lake, it had commercial fish docks, a hotel, grocery stores, barber shops, and other businesses to accommodate the public's rising interest and needs. It seemed to have its beginning about 1898 when the Reelfoot Outing Club was established. The first lodges had been built, sure indicators that the lake was destined to be a popular hunting and fishing destination. By 1915, there were at least three club houses at the lake, clubs that usually had "sports" to guide, and Lakers had the skills to accommodate them. The Covington Reelfoot Club, according to R. E. Lee's book *Reelfoot Lake Fishing and Duck Shooting* (1915), was said to be the largest, which had sixty members from Covington, Tennessee. The club sat on three acres they owned on the south shore of the lake, still known as Blue Bank. Reelfoot was a lively place during those years, when there were often not enough guides to accommodate visitors to the lake.

Just beyond Walnut Log, on Grassy Island, the road ended at the Blue Wing Club. The clubhouse was eventually moved to Gray's Camp, where it stands today. The old Black Jack clubhouse, owned by the state and leased to a club, is also located here. At the former site of the Blue Wing club house is a large parking area, a boat ramp, and a US Fish and Wildlife Service boardwalk with a viewing tower; it had barely enough opening in the tree canopy for a tent camp back in those days. The road there was a rutted wagon road in my day, and the site was known as "Blue Wing." Today there is a paved road on Grassy Island to Blue Wing and the rest is wilderness.

The country was on the move during the 1940s and 50s, especially among the younger generation, and our remote community of Walnut Log felt the change –a drift from remote living to the new styles and conveniences of urban life. But it seemed to affect my folks and others like them very little—they still lived the Golden Age, and decided to stay in their home communities and make the best of life. Move from the lake? I don't remember the subject being discussed until Dad decided to farm. It was much the same in all of the Laker communities around the lake. It could have been better only if we had lived all of the "Golden Years" our fathers described. They were the rubber-booted guides, trappers, commercial fishermen, woodsmen, and craftsmen I wanted to be. They seemed to think that because they had lost black bears, cougars, elk, and even deer and wild turkeys, things were really going downhill. But they were blessed with comforts that David Crockett and his ilk never knew. I could accept that; I just hoped the things we had in the present would always be here.

The best years for Gumbooters might have begun immediately after the Civil War and lasted until the mid-1950s. It was a happy time for most; the latter years were the easiest to remember. The entire family was generally involved in making a living and cheerful about it, mainly in the arts of fishing, hunting, herb-gathering, and trapping. Cotton chopping (zero cheer); cotton and tomato picking (not so bad) had immediate feedback; the harder you worked, the more money you made. It is true that during the earliest of those years we had few modern conveniences. I remember Mom's longing for electricity, a washing machine, a refrigerator, or vacuum cleaner, but only mildly complaining about it. The change really did make life more pleasant for moms. I suppose it did for the rest of us, but I was a slow learner in these subjects and preoccupied with things that had little to do with housekeeping, so I know could not appreciate how important they were to our family.

But those times improved, little by little. By the time I was eight or ten, we had electricity instead of kerosene lamps, a refrigerator instead of an icebox, a window fan instead of open screened windows, and a few other things, like gum boots of the right sizes. As soon as kids outgrew knee boots (usually castaway, cut-off hip boots), they wore this hip-to-toe gumboot footwear, labeled "Red Ball." The boots were usually worn casually with the upper part rolled down below the knees. If the choice was between long pants or these boots, the boots would likely win. That's the way it was during those years—the years I was fortunate to live in the remote community of Walnut Log.

In our house was Mom and Dad, a sibling brother and two sisters until I was a teenager; then, there were two more sisters. It was quite a life for a kid—where else in Tennessee could you find footpaths along a bayou, stump-jumper boats parked within fifteen steps of the porch, frogs croaking, and fish jumping day and night to calm our attention for sleep? Only quarter mile upstream from the Walnut Log Canal and McQueen's Grocery was the Upper Canal, something of a secret passage to the very north end of Upper Blue Basin. Both were on the east bank of the Bayou du Chein, and so was the footpath to Aunt Ivy and Uncle Brady Henson's house a little farther up the bayou. In between was a number houses with kids like me, along with a lot of opportunities for adventure.

Yes, it was a life of wonderful memories. Crossing boardwalks and romping foot paths barefooted with a fishing cane on our shoulders was as common a sight as white "cranes" fishing from logs. But you will not see a single house across the bayou on the west side today. As mention elsewhere in the book, those houses were removed by the state in 1986. Its census did not increase much beyond the heydays of the 1940s and 50s, nor did its businesses survive, mainly from change of non-resident ownership and lack of good business sense. For those who could not live without wild things around them, who needed little more than honest neighbors, a good boat, a little gear, free time, and a warm, furnished cabin, Walnut Log was the ticket for a good place to live.

Guides and commercial fishermen were always moving about along the way; they had news and stories about things and creatures that were not only educational but also appealed to a kid. It was a time when the vitality of the lake and its people were at their very best. It was years before I realized that I had lived part of "The Golden Age" at Reelfoot Lake—and I, too, was raised a "Son of the Swamps."

We were a homespun family. Walnut Log families often started their families as young parents, not so unlike most frontier families that needed children to grow up as an integrated team with strong backs and spirits to help the family make a living. Rarely did we wear anything store-bought or new; nearly all of our clothes were homemade cotton or hand-me-down clothes from neighbors or cousins. Of course, sometimes the oldest kid had the privilege of being the first owner. And, like me, they rarely wore shoes until the ground began to freeze. It leaves one to wonder—in the absence of thermal and goose-down winter clothes, how did we survive the exposures of an outdoor life during those cold, cold winters? Our heat source out in the cold, especially duck blinds, was a bucket of glowing charcoal. I am often surprised we ended up with toes. We managed, but there was little pleasure in being "rotisserized" to keep from freezing. Still, I don't remember anyone leaving the lake just because they might freeze to death.

Few Laker children of my generation wandered far, or didn't return once they grew up; they sowed their wild oats, reflected on the lure of home, and eventually returned or wished they could. To move farther than a good walking distance from our parents and the bayou didn't make sense when we might not own an automobile. The best mode of travel along our routes was a wooden bateau. The songs of the lake sirens were strong; there was something about the lake I've yet been able to fully describe. It spoke to one's soul about peace and comfort, a place where one should be. It took a dire circumstance to leave such a place, let alone to be far from the warmth of one's kith and kin.

Walnut Log Gumbooters were typical in that they didn't fuss too much about floods, even before the levees. For the most part, new levees along the Mississippi were adequate for all but the worst floods; at least that was the feeling. Even if floodwaters entered their houses, which was common enough, families took it in stride. Our family never moved out of the house. Shoes were placed on a shelf, the kitchen stove was still operational, and beds were kept dry by putting the frame on blocks. It was something of a fun novelty, speaking as a kid, to step out of the bed the next morning in ankle-deep water. I don't remember Mom being too upset; she simply rounded up my brother, sisters, and me and assigned duties. As soon as the water receded in a week or so, the mess was cleaned up, things put back in order, and life went on as usual. Dad was usually out on the lake taking advantage of the heavy runs of fish that followed floods, until he was called to war.

Floods, I'm sure, were secretly worrisome for Mom, and all moms around the lake, but you'd hardly recognize it. Not so much for men. Floods were also a time for opportunity. The lake teemed with fish and wildlife during floods, and the men spent much of their time commercial fishing or hunting. Though sympathetic that a flood was troublesome, they knew its importance—fish moved during these spells and could fill their nets. It was common sense that floods rejuvenated the entire lake ecosystem. After all, the men had seen floods all their lives; the slowest wit could see that natural rivers flooded and dried their floodplains with wet and dry seasons; natural as breathing. It was easy enough to see that fish followed fresh river water brought to the lake. There were reasons: fish and wildlife food was available, and the spawning season was during spring floods. Flooded fields and forests meant fresh food in abundance; fish have more sense that you might think; they had an advanced intuition, and so did smart fishermen and wildlife predators that lived on fish. Fishing, hunting, trapping, and everything they treasured about the lake was better because freshwater brought life during floods.

Figure 12. Floodwater was three feet higher than normal:
summer pool during local floods of 2011 and 2019.

Spring floods are necessary for fish populations. Something that might not be commonly known is that most fish with a notion to spawn and raise their broods also need access to remote sloughs, ponds, and marshes with a favorable habitat to lay their eggs and raise their offspring. In this protective environment, youngster fish grow strong enough to leave with a single-minded objective: to replenish lakes and rivers with their kind. Understanding the world we lived in helped lessen a string of troubles or inconveniences caused by events like high water. An axiom in our family was fairly common to most Lakers: nature does not make mistakes; the sooner we accept her, the less disturbing her consequence affects us.

It was far from the life of urban dwellers. Troubles sometimes encroached upon the lives of Gumbooters, but most were soon resolved. By the second decade of the 1900s, early growing pangs had gradually passed, and the outside world had been smitten by outdoor life at the lake. They came, and some stayed. One would be hard pressed to find anyone who'd lived at the lake for any length of time, who did not desire to spend the rest of their life here. So it was pretty much this way for as long as I can remember.

The commercial fishing business had its own interesting history. J. C. Burdick's commercial fish business of the late 1800s soon expanded at Walnut Log to include a few facilities for lodging, dining, and guide services to accommodate sportsmen. His business mind could see the rising interest in the lake's extraordinary hunting and fishing sports, an opportunity not to be ignored. Marvin Hayes of my day was the guru of commercial fishing. Burdick was known as the "King Commercial Fisherman." Hayes was also the "Godfather" of commercial fishermen. He had the only commercial fish dock at Walnut Log for as long as I can remember. But while commercial fishing remained the first major breadwinner at the lake for many years, Burdick's business radar was working just fine; sport hunting and fishing were top billing by the early 1900s. Even though visitors suffered the early rough gravel, and often muddy dirt roads from the towns of Hickman and Union City, wholesale buyers and sportsmen kept coming. With better roads and more room to expand, Reelfoot Lake commerce was on a roll.

Pressure on commercial fishermen had already begun before 1908 with all the ballyhoo about lake ownership and fishing rights, but some of it continued for another half century. Other than who had the right to fish the lake, the commercial catching of gamefish was always the issue. Sizeable commercial catches of largemouth bass, crappie, and bluegills were too much for a growing population of sport fishermen. Poor catches and smaller fish by sport fishermen seemed always pointed to commercial fishermen.

In a letter of April 17, 1930, shown to me by Helen Carringan (daughter of Steven and Lillian Crossley), Walnut Log Lodge owner O. T. Wallaston, during the 1930s, had this to say to Steven Crossley, the lake's first game warden: "Commercial fishing now is about played out. I think the natives are about to realize that a greater income may be derived from the (opportunity) of sportsmen than the sale of fish today."

The lake of the 1930s was already leaning toward sportsmen recreation as the major commerce. While commercial fishing had set the character of Reelfoot as sure as duck hunting and sport fishing, it would soon be history. But the warning by O. T. Wallaston was hardly mentioned for another twenty-five or thirty years. Commercial fishing remained the recognized bread-and-butter enterprise from the days of J. C. Butdick Lake until about the mid-1900s. It was a very active enterprise, with 150 or so commercial fishermen, and one that was very much appreciated because it meant we were never without fish to eat. Burdick's and Marvin Hayes's fish docks at Walnut Log and Samburg were probably the primary business enterprises at the lake well into the 1950s. Not only did these fishermen sell fish and turtles for the market, but these were staples for family table fare; fish that didn't make the market (called "culls") made the family dinner menu. A platter of fried crappie on the menu at local restaurants was also a widely popular item. We did not filet small fish, but scaled, and fried them whole—not that moms didn't bake fish, particularly buffalo and carp. Most of us preferred cull crappie for many years. Big "slab" crappie seemed coarse and too much meat for one bite; smaller ones were sweet, fine-textured meat, and cornmeal-crispy.

The real problems began when all game fish but yellow bass (stripes) were made illegal in the early 1950s. In 1953, articles in local newspapers stated how bad fishing had become; more and more of the sport fishing public demanded that commercial fishing be stopped. The Tennessee Game and Fish Commission were compelled to conduct studies in order to evaluate the matter. The first complete census of commercial harvest was conducted in 1954–55. A total of 644,911 pounds of fish were harvested by 150 commercial fishermen. The total included whopping 75,753 pounds of crappie, 64,838 pounds of sunfish, 272,225 pounds of catfish, and 193,764 pounds of rough fish (mainly buffalo and German carp).

The Game and Fish Commission study had a tremendous influence on the future of commercial fishing at the lake, Reelfoot Lake was the only lake in Tennessee in which game fish (largemouth bass, crappie, and sunfish—mainly bluegills) could be taken and sold commercially. Strangely enough, the six-year interval of closures showed that the average growth rate and weight of white crappie, the esteemed sport fish, actually declined.[3]

The commercial sale of crappie was revived in 1964 and stayed that way for another thirty-seven years. But for nearly half a century, the commercial taking of game fish was a hot issue between sport fishermen and commercial fishermen. Ultimately, sport fishermen had their way; commercial harvest was finally made illegal for all sport fish. This was the "straw that broke the back" of commercial fishing at Reelfoot Lake. The legal catch of crappie, bluegill, and bass by commercial fishermen was the "break-even" part of their daily catch; the rest of the catch was profit. It kept the commercial fisherman from becoming a relic. Markets for rough fish such as catfish, buffalo, and carp were not enough to sustain their business.

Most commercial fishermen today do their work for a single reason: it's "in their blood." They do it because their former occupation was their heart and soul. If you happen to find any commercial gear on Reelfoot today, more than likely it is set to catch catfish, maybe to sell, but probably for the pleasure of their own use, or to provide a mess for their neighbor.

Figure 13. "Boochie" Berry's set net; commercial fisherman and Reelfoot historian.

Commercial nets are the only practical way for most to catch buffalo and carp for table fare.

Once numbering 150 or so registered fishermen, their numbers dwindled to two dozen as their occupation neared the final days of an era; it was only the beginning of more regulations and contests unfavorable to commercial fishermen who could not easily leave the only life they knew. For reasons I can't yet describe, fried crappie in Tennessee seems to have the same high esteem as halibut-on-the-plank in Alaska. For sure, crappies are a culinary treat for those with a palate for fish. With fried okra, French fries, hushpuppies, or twice-baked potatoes and a tall glass of ice tea, nothing else is needed on the menu. Crappie fishermen will tell you that the best crappie they ever had was the last time they had fried crappie. Yes, indeed, but insofar as I am concerned, the same can be said for fried bluegills. And some I know prefer fiddler catfish filets. At any rate, it was not unusual during the thirty-seven year grace period that crappie could be commercially harvested and sold (since 1964) that visitors commonly drove a hundred miles one way to dine at one of Reelfoot Lake's famous restaurants. I have little doubt that many sport fishermen had mixed feelings about removing crappie from the menus when they had a skimpy fishing season.

Trying to accommodate the needs of people and natural resources at the same time is fickle business for TWRA. Just to show how fickle, I offer the following: Based on the Game and Fish Commission's study to determine the effects of commercial fishing on the crappie population, the study showed no great detriment to the crappie population. So, the commercial harvest of crappie was renewed in 1964. Indeed, but it was under strict control. Every crappie had to be brought to a check station, weighed, measured, and tagged. In addition, only limited poundage of crappie could be commercially harvested. But, alas, this was not satisfying to sport fishermen—it took thirty-seven years, but a prohibition against the commercial harvest of crappie and game fish, I believe, was final in June 2001.

Commercial rough fish today is usually limited to catfish, German carp, black buffalo, and yellow bass (stripped jacks). Asian carp are usually considered a menace, but some they are good table fare. Commercial fishing today is a relic of the past. We caught only two kinds of fish for sport when I was a kid: blue gills and largemouth bass. As I have mentioned, crappies, much to a family's advantage, were commercial fish. Families enjoyed the "culls," rarely saving the larger "slab" crappie since these were market fish.

Figure 14. Wendel Morris and Terry Pride brave the cold to set trammel nets.

The old tradition of commercial fishing is a fading art. Sometimes old-time fishermen still haul out their stained and gill nets laden with spider webs and dirt dabber nests to catch a few buffalo fish to eat, and to share with neighbors and friends. Buffalo and winter carp were choice table fare for us years ago. The large rib sections of buffalo or German carp, as I mentioned earlier, have no small bones. Caught when the water is cold, the large rib sections of these fish—fried nice and crisp—make robust and delicious main-course meals.

Controversy during the 1950s finally resulted in laws passed to outlaw wire baskets and other kinds of commercial gear. It was yet another problem that compromised liberal rules for fishing at Reelfoot Lake. Right or wrong, it dismissed an ancient livelihood honed as an art and science by skilled, hardworking frontiersmen. Some of it was self-taught, but most of it was a century of fishing passed on by kith and kin.

Reelfoot Lake was still a wilderness during the mid-1800s; even as the lumber industry thrived to supply our fatherlands and a growing nation, it had a virgin forest. Can you imagine the news that a virgin forest lay hidden at Reelfoot Lake? It rang like the jingle from bags of gold in the ears of the lumber industry! Since Reelfoot was considered a public wilderness by locals, cutting down trees for whatever use they chose raised no eyebrows. There were no restrictions until the early 1900s, when some like J. C. and Judge Harris claimed certain land titles around and beneath the waters of the lake. Like many progressive endeavors, once lumber companies saw the potential, the industry blossomed; they wanted it all, and they wanted it now—and they had no interest in a little help from professional foresters. There was a way to do it, not with pseudo-progress, but real progress. Conservation and sustained yield was not a popular subject in those days. But sustained yield and good forest stewardship is still the future for the forests of Reelfoot. We can do that with all of our natural resources, like the wetlands of Reelfoot Lake. The way the lumber companies (and a progressive nation) went about it was like taking only the tongue from a fresh bison kill—take the best and waste the rest. What we idly seem to ignore is that this lonely planet in a universe with billions more is the only one fit for humans. We cannot afford to thoughtlessly disassemble great landscape diversities like Reelfoot Lake, one natural resource at a time, until it's too late to reconsider and reverse the trend. It is the four inches in depth of nutrient-rich topsoil sustains us. This is where the richness and diversity of our landscape begins; like forest lumber companies, we hope to pass the forests on to the next generation.

My grandfathers were guides ("pushers") and commercial fishermen, but at intervals they were lumbermen. A hazardous business, harvesting timber had cost one his left eye. Most of the forest had not seen the bit of an ax or the teeth of a saw until about the mid-1800s. Giant cypress, oaks, hickory, poplar, and walnut were prime trees that grew in cypress swamps, on shallow ridges and the upland around Reelfoot Lake. These floodplain trees were among the fastest growing with some of the highest-quality lumber in the world. The diameter of virgin cypress logs often well exceeded the height of an average man. One sixteen-foot log was sometimes all a team of four mules and a wagon could haul.

Not even the deep swamps of Reelfoot Lake offered security to virgin forests sought by hungry lumber markets; although ingenuity was needed to snake out these great logs, local sawmills thrived during the 1870s through the early 1900s. You might imagine trying to skid huge logs out of the swamps on soggy corduroy roads made of logs and poles. The lumbermen often chose to wait on high water to float out their logs. TWRA often found well-preserved but abandoned logs sunk in the mud when they cleaned out canals and streams. You will still find the large moss-covered stumps of these harvested titans while quietly passing through on a carpet on cypress leaves any of several old-growth cypress swamp.

The future of the forests at Reelfoot provides an excellent opportunity to address the future of the lake itself. Between the 1860s and 1870s, a half-dozen or so sawmills were known to operate in the vicinity of Reelfoot

Lake. Nearly all of the forest around the lake considered not too wet to farm had been cutover by the early 1900s. There was little thought that a sustainable forest could one day be more valuable than cropland and eroded hillsides. Forest management, like wildlife management, is a recent profession. Both were dimly acknowledged before Aldo Leopold, Father of Wildlife Management, established them as science, and brought the professions to the public's attention during the 1930s and 40s. It would be another three or four decades before the subjects were taught as majors in Tennessee colleges and universities. So forest and wildlife management is still a very young science; the science of restoring native wetlands has just begun.

You might be surprised to know that the mandate to maintain the integrity of Reelfoot Lake has not passed us by without notice. In the 1980s and 90s, volumes of research, investigations, meetings, environmental assessments, and master plans were completed; they lie dormant in files today. An all-out effort and millions of dollars were spent in an attempt by state and federal agencies to get the management of Reelfoot Lake on track for the future. You hear nothing about it today. Perhaps it will be useful in the future.

The lumbering industry reached its peak and was virtually over as an industry within fifty years. It gave more incentive to simply clear what remained of the forest for cropland, often on hillsides overlooking Reelfoot Lake that were much too steep to farm. The soil frequently became so depleted on these erosion-prone fields that they too were abandoned. Bottomland hardwood forests were no exception. In fact, most of the forests in the Mississippi River floodplain were cleared and farmed. The remnants of the forest were marginal wetlands considered too wet to farm, and these lasted only a short while longer. Now we can reflect on the great wealth, economic and biological advantages our forefathers took for granted during the nation's early quest for progress. So we can see the setting around Reelfoot Lake as a microcosm that pretty much replicates the history of wetlands from one end of the great Mississippi River Valley to the other. By the 1960s and 1970s, even these few forested acres were gone: Not even a row of trees for wind breaks remained, nor were there wooded buffers along streams, a corner acre of wetland for a rabbit or quail, or anymore native wetland to clear and farm—all of it was gone. Cleared wetlands went to a short-season crop—soybeans. With the once great diversity and distribution of forests and common plant habitat gone, gone also are the tiny creatures that support the food web for fish, fowl, mammals, and all other animal life.

Soybeans finally replaced practically every sprig of former fallow ground in the river floodplain, and the last acre of fast-growing, high-quality bottomland forests on private land. The agriculture industry said we needed those acres "to feed a hungry world." We all know how important the agriculture industry is to the wellbeing of this country, but how will our hungry world cope now that those acres are all used up? We are compelled to ask, was there not a better way to equitably share and manage the use of this great natural resource? There will be others after us who will need more than soybeans for a quality life, and there is no indication that any of it is found in the outer universe.

How do we recover the benefits lost? The task is enormous when considering the ebbing away of the world's greatest bird flyway, the world's highest-quality and fastest-growing bottomland hardwoods, the filtration zone of sediments and pollutants entering the river, the sequestration of industrial carbons through forest photosynthesis to cleanse the air, the decline of a sponge to capture groundwater, and so on? This thing we casually call "progress" at the expense of a massive loss of natural resource diversity may be more insidious than the careless use of DDT; our silent springs could last the summer, fall, and winter—all seasons, and a good bit more.

We still have a remnant of the virgin cypress swamps at the lake. Similar stands of trees are very rare in West Tennessee. A little wilderness adventure will help you find these isolated patches on the WMAs and refuges. Relics of sawmills and railroads in the western forests of the Reelfoot near Little Ronaldson can also be found. Barges transported logs from these swamps to the southeastern side of the lake, via trails cut through stump fields of Lower Blue Basin. From here, the logs were processed at local sawmills or shuttled to railroad cars for transport to other places.

While most accessible trees were harvested, a few aged stalwarts were just too deep in the swamps to be conveniently reached. The sawyer might have wondered as well when he pondered the meaning of the growth rings at the butt end of a giant tree he had fallen: What history would the growth rings tell him; each ring a year in the book of history; five-hundred or maybe more, now in the sawdust piled at his feet. But a thoughtful sawyer might have counted the growth rings; ninety years back was the birth of his great grandfather; the tree was a sprout when Capitan John Smith's Powhatan princess Pocahontas had turned sixteen. So, many of these trees were titans when Henry Rutherford and his crew floated down Reelfoot River and set their first survey stake. Colonel David Crockett walked beneath them in pursuit of black bears. Fortunate for us, many remain untouched for us to ponder and wonder: what man or beast passed beneath the canopies of these giants since they were mere bushes?

Once at the lake, the best way for early visitors to get around or through this vast wetland was by boat because a circuitous road around the lake was still decades in the future. Between Gum Point, just south of Samburg, and White's Landing (formerly Gooch's) was a four-mile stretch cleared of underwater stumps across the south end of the lake, mainly in the interest of the Keystone Lumber Company. A fleet of manpowered skiffs, stump-jumpers, and a tour boat or two uses these routes. But benefits to the lumber companies didn't last long; word was passed along that their steam-powered craft ran over a stump and sank in Lower Blue Basin. Its greatest benefit might have been to savvy commercial trammel net fishermen who knew where the boat sank, and that it was known as a good fish bed.

This brief look at the history of the lumber industry parallels the history of the lake itself; as you have seen, over-exploitation and mismanagement can ruin the ecological integrity and potential for natural resources to serve our needs for the future. Cutover forests have begun to recover, and managers today are looking at Reelfoot's forest as a rare remnant of bottomland hardwoods, with no harvests planned in the near future. Because old-growth bottomland forests are so uncommon, the remaining forests will probably be managed as an old-growth forest, where trees with cavities will be large enough to den black bears. Since the primary purpose of this forest is for wildlife, managers will probably allow nature to take its course; even natural fires and storms (and even earthquakes) are part of nature's scheme to diversify and enrich the woodlands. This forest will once again produce an ecosystem favorable to native plants and animals that made up our wildlands centuries ago. See what you think as you hike the trails, meander through the open forests, canoe the canals, and cruise its backroads, and consider the history.

Changes and Adaptations

Beyond the loss of a commercial fishery, making an easy living at the lake with old methods of the past is even more difficult. More restrictive game and fish laws in response to more public demands and use is a reason, but social mores is another. Modern trends like television changed how local school kids looked at the world; lifestyles and industry drifted from the old norms, and the younger generation found these temptations more abundant in the big city. Some began to think commercial fishing, guiding, and trapping were too much like work with too little income.

Still, a few since the mid-1900s had a sixth sense that the world outside their boundaries could not improve their lives. Among them was a remnant that had the same high esteem for the lake as Mr. Lexie Leonard, Mr. Bruce McQueen, Mr. Wendell Morris, Mr. "Boochie" Berry, and a good many others; and I was one. Some of these still homesteaded on public land well into the 1980s. The ancient tradition of claimed territory as theirs to hunt was remained a policy by many who trapped, commercial fished, or hunted waterfowl. Yet, with the passing of time, more and more changes were instituted by people than by nature—more developments, more land clearing, more faith in the fragile security of levees along the Mississippi; we began to lose the reality of living in the floodplain—and with it, the awareness and value of nature's lessons our fathers knew so well.

So, the lifestyle of the Golden Age for most Gumbooters has passed. Slowly, but surely, the old ways of making a living have faded as the trend to the age of outdoor tourism demands more attention. But it is only a new age with new opportunities and new ideas, and it is likely the future as values change and the lake's ecology changes with age (or lack of good care). Hiking boots, trails, and boardwalks could be as popular as push-poles and stump-jumpers. It might be tennis shoes and kayaks and fewer airboats and slick bass boats. It would be a good thing as well for the coming trend of outdoorsmen. But sportsmen have held the conservation banner high and have been the major financial source of fish and wildlife programs for a century. The future will require more and more assistance from non-consumptive users with a friendly attitude and a passion for nature. A new generation of Gumbooters will see this as years go by.

Reelfoot Lake remains, for those who love nature, a rare jewel, an oasis for weary wildlife migrants, and a natural garden retreat that will appeal to many for years in the future. Like nature itself, the world changes; the old and the restless new must constantly adapt. Efficiency is in the name of nature, and she will continue to seek easier and more efficient ways to survive in harmony with the changing world around her. With wise conservation management, her glory can be ours to enjoy for many generations to come.

Chapter 6

The Life of a Kid on the Bayou

For a Gumbooter kid, life at Reelfoot Lake could not have been better suited. Until the mid-1900s, normal life at the lake depended heavily on the natural resources of the lake. I don't remember a single dull day in my early life here. The lake was a society in which about anyone with a decent work ethic could make a reasonable living from the lake's natural resources. It was a society that loved the lake as it did its country. Few would say they lived a dull life. There were things to be concerned about on the bayou, but the world we lived in was simpler then, and ultimately the place of my earliest memories are where most of the stories in this book begin.

Home through my teenage years, the woodland hamlet of Walnut Log was on the Bayou du Chein, which nurtured Reelfoot Lake. The citizenry of Walnut Log was close-knit. The head of the house and most of the family depended upon hunting, fishing, and guiding—generally living off the land. Neighbors were mostly considered family and, in fact, several were; grandparents, aunts, uncles, and cousins—many lived along the bayou or nearby. Neighbors kept an eye on the young; a kid born and raised here was about as free as a bird by age three or four to roam almost at will. Scolding was usually mild when rules were broken, unless the youngster wandered on the docks and bridges of the bayou before age four or five. Shy around strangers, children hardly knew the outside world before local schools were consolidated.

Any outdoors person would love this life. The community of only two dozen cabins and cottage-like family houses was tailored for those who love to live in the outback—warm, dry, plenty to eat, and no rent. About half of the homes were on one side of the bayou and half on the other. The west side of the community was the most remote because it was a strip of high ground across the bayou, accessed by boat or boardwalk bridges. It was also the favorite playground for adventurous kids. This, of course, was where my family lived. Mom and Dad raised three younger siblings and me here in a two-room cabin with a small kitchen and a screened front porch. Sturdy cypress boards were the exterior, a weather barrier of tar paper behind it, followed by decorated wall paper on the inside. The roof was cypress-board shingles, hand-rived, with an expected life of about one-hundred years.

Visitors were welcome. Distant visitors often made a cruise through Walnut Log as part of their scheduled trip to the lake; locals, for a Sunday drive. Many only wished they had the good fortunate to live here. "How did they survive," others whispered, "in this snake infested swamp?" A concern not all together unwarranted. At least there were no alligators. Not often you could find a community where children played relatively unsupervised along wooded footpaths, at the edge of a swamps and bayous. Simple: our swamps and bayous were analogous to an urban backyard. A snapshot of my youth would be a picture of a kid traipsing along the barren footpaths of the bayou with a switch-cane fishing pole on his shoulder and a bucket in hand. Dangerous? Never thought about it. I don't remember a single life-threatening tragedy for a kid raised in this woodland life, although there were times when I wondered how we avoided it.

Foot paths were our sidewalks; bayous and canals, our highways. Damp and barren footpaths were well-suited for summertime kids and bare feet. There was a labyrinth of them: some ran through the woods behind the row of houses, some onto the marshes a few hundred yards outback, where a boat might be parked at the edge of the giant cutgrass as access to the lake. Others went to outbuildings, the community store, and other places. But the main footpath followed the west bank of the bayou, since there were no streets here. It ran from the Upper Canal a quarter-mile downstream to the Walnut Log Canal and McQueen's Grocery. This was the end of the community. Along this path, my friends and I spent many summer days fishing and poking for creatures in the bayou.

Our father was not around for a couple of years; he was a WWII soldier at the Battle of the Bulge. His Reelfoot stump-jumper, built and customized specifically for Reelfoot, was still parked in the boatshed where he last parked it. And there were empty cypress-boarded fish and turtle live-boxes, still waiting for fresh catches. His absence put life as usual on hold—except for the kids; we were precocious enough at a very young age, already acclimated to this life. Meanwhile, Dad's absence made Mom the undisputed captain of the house. I had not yet started going to school, so the world on the bayou was mine to explore for some years—as long as Mom approved, or was blessed by being oblivious to my explorations.

Daily life was often with my troop of friends, a small knot of adventure-seeking "swamp rats," as we were frequently called, all of a similar age. That friendship lasted through our teenage years, or until someone moved. Years later I read Mark Twain's *Adventures of Huckleberry Finn*. Twain's characters were from Hannibal, not far across the Mississippi. The novel reminded me of a kid's life on the Bayou du Chein. There were several in my group of characters: Spencer Powell, Tobe McQueen, Ted and James Carter, and my cousins, the Henson boys, as starters—all of the same ilk and fitting for a chapter in a Mark Twain novel. Brave sisters might join us, but their participation was usually tentative. They didn't care much for war with a nest of angry bumble bees or yellow jackets. Collecting green snakes, baby turtles, and cricket frogs, or cleaning out a bee's nest with wooden paddles substituted for urban sports we hardly knew.

And what would a gang of young swamp rats be without a dog? Well, there was Spot—yes, Spot, our pal, my cousin David's fearless dog. Spot was our mascot, a medium-sized dog with a Heinz 57 pedigree, a white coat with a black patch over his right eye, and one on his left side. He would play sock ball (we had no regulation ball diamonds), dig out a bumble bee nest, reduce a provocative cottonmouth to smithereens, or tackle a chicken-eating bobcat. He was always on our side. Fictitious? Nope; just a typical day with Reelfoot swamp rats.

I'm sure we were often a worry for moms. Busy caring for my younger siblings, my mom often depended on neighbors to help assure that the swamp had not consumed us. Mosquitoes? Mom fretted more about mosquitoes than cottonmouths. Malarial; I heard that word a lot. We were reminded that the disease was caused by a mosquito bite, delivered through the hypodermic proboscis of a female anopheles mosquito. Large spray tanks on the back of trucks full of insecticides coated the community every summer with acrid-smelling spray. Malarial (the Plasmodium disease) must have been a dreaded thing, although I knew of no one with that infirmity. Yet, it was often suspected when one complained of the slightest fever.

One thing we did know was that the despised, high-pitched "z-z-z-z-zinging" from the wings of a blood-thirsty mosquito was a summertime pest. It could ruin a night of sound sleep. Mom had remedy for that. It was a nightly ritual at bedtime during the summer: "Cover your heads

Figure 15. The Johnson Family on the Bayou du Chein.

children!" And in she would come with her DDT pump-gun. No mosquito in its right mind was safe at our house. With the canister brim full of the deadly propellant, her mission was to exterminate every blasted one. Those that didn't die immediately were chased down with a fly swatter.

No matter how much cover we put over our heads, we could not hide from the breath-stopping stench of bug spray. When it came to Mom protecting her children, not even a Bengal tiger stood a chance; but DDT? Little was known about it. The government said it was the pesticide of choice—the vector for Plasmodium had to be eradicated. But it was questionable who would croak first—us or the tormenting mosquitoes.

Walnut Log Community Life

Like the rest of the society at the lake, Walnut Log has changed. Unlike the lively community we found when our family lived here, most of the resident houses, two stores, a fine lodge, and the Tennessee Academy of Science Biological Station are gone, their yards grown up in trees. Footpaths are deer trails. All that remains are a few houses on the east side of the bayou, mainly along Bennett Johnson Road. The population is less than a third of what it was at its peak, and the lively pack of Huckleberry Finn "swamp kids" are not to be found. But let your imagination wander. Maybe this account will invoke the ghosts of those days.

Walnut Log buzzed with activity year-round well into the 1950s. Commercial fishing and hunting were top billing fall through early spring;

sport fishing, gardening, and other activities kept us busy during summer. Guides for hunters or fishermen were in demand the entire year. Lives that depended mainly on the goodwill of nature were not disappointed. A gravel road to Walnut Log ended at Walnut Log Lodge, the center of community activity, the engine that drove the tourist, fishing, and hunting industry. Walnut Log Canal was access to Upper Blue Basin. From the front porch of Walnut Log Lodge and the boat launch, it was a straight shot to the open lake. The boat launch, bridge, and dock and fish-cleaning station were on the bank of the bayou. Sports cleaned a train load of fish and told a sizeable hill of hunting and fishing stories at this dock. McQueen's Grocery, just across the Bayou du Chein, on the "roadless" side of the community, served the same purpose.

Community life still possessed a flavor of yesteryear's simplicity—boat houses along the bayou with stump-jumpers and skiffs; kids laughing; live-boxes with fish and turtles; nets hung to dry; a Hank Williams song on a radio. Paved roads are a relatively new thing; roads through the community are surfaced from local gravel pits or dirt. Well into the late 1950s, nearly all roads around the lake were gumbo dirt or gravel. The crackle of tires on gravel could be heard more than a mile away when a vehicle was coming or going. All Walnut Log lodges were thriving businesses through these years. In my day, the site of Walnut Log Lodge was Ward's Lodge in 1908, renowned more for its role in the infamous Night Rider days than for anything else. Ward's Lodge was unique in that its foundation rested on large cypress logs that floated the lodge at high water. Sometime around 1918, the lodge was replaced by a large cypress-boarded facility with a solid foundation of eight-foot piers. Owned by Mr. T. O. Walston, every succeeding lodge has been known as Walnut Log Lodge.

Pause at the boat ramp and imagine for a few minutes a summer day here during the late 1940s, about sundown. You find a deep feeling of calm and peace settles as a not-to-be-forgotten memory. Watch the soft glow of a red sunset settle low over the lake and hear the quiet "click-clack" of oars coming from the canal as the last fishermen in stump-jumpers return from a day of fishing, or a pleasure boat ride. Locals begin to gather across the bayou at McQueen's Grocery or at the lodge fish-cleaning house, a convenient gathering place for Lakers both to find out how fish were biting and to catch up on local news. Those at McQueen's Grocery were often accompanied by their youngest children, usually found enjoying a cola or a sweet snack. Listen closely and you might hear the murmuring of their low conversations. You might notice a knot of barefooted kids coming down the footpaths, some with a fishing cane over their shoulder and a tin of bait. You watch as they clamber to get the best position on the Grover McQueen's floating dock for a last minute of fishing.

Hear the close of a screened door. It's behind you on the porch at Walnut Log Lodge at the end of the gravel road. A mother somewhere up the bayou calls to her children in her best high soprano—Supper! It could have been Mom. You are reminded that the scurrying about of village life here was not so unlike that of a Cajun bayou community some six-hundred miles south.

Walnut Log Lodge rose from the earth on tall piers like a living, green dome. It was actually a large cypress board-and-batten building with a dozen or so guest rooms and a large lobby with vaulted ceilings. The outside was painted park-green, very much in the flavor of National Park buildings of that period. A wide set of twelve cypress plank steps stopped at a long screened porch. The porch looked conveniently straight west through a gap in a small grove of giant cypress trees on the bank of the bayou. At the end was a view of Upper Blue Basin. Today, you will only see barren ground where the old Walnut Log Lodge sat. The old lodge left many joyful memories and housed thousands of sportsmen and other visitors for half a century.

The lodge burned in the 1960s. The center of community celebrations and lodging for guests from all over the country was finally history (later it was replaced by a red brick motel that might have fit the suburbs of Chicago, but not Reelfoot. It too burned, and was the last lodge at Walnut Log). I was on the backup guide list for the lodge, and was called fairly often, school or no school. The lodge was the pulse of the community during prime waterfowl and fishing season. Business generally went on as usual during high water. High water was so common, people thought little about it. Some considered it quite convenient, in fact, to pull all the way up to the lodge's lofty steps by boat in river water.

Mississippi River floods? Not a problem. Since the old lodge rested on eight-foot cypress piers, the lodging quarters were well above excessive spring floods. Beneath the lodge were wire-caged rooms for storage. Automobiles sometimes parked in the free spaces. A more fitting architecture for Reelfoot or a better hunting and fishing lodge would have been difficult to find.

J. C. Burdick (known as "Mr. Commercial Kingfisher") is said to be the sponsor of the Walnut Log Canal, which was rebuilt by the state in 1947. I think of Mr. Burdick as more or less the Father of Walnut Log, as it was

Figure 16. Walnut Log boardwalks across the Bayou du Chein.

his commercial fish business that defined and carried the name of the community. Burdick moved his business to Samburg in 1870.

Cypress board and batten were the materials to set the character of buildings at the lake for more than half a century. Fifty or so residents were about the peak population of the community, and their houses were usually built on piers, until the river levees were constructed. But with as many as two dozen residence and two small stores when I was a kid, only an older house, the lodge, and the Tennessee Academy of Science Biological Station were on piers. Characteristically, houses were cypress board and batten, with two or three rooms, a potbellied stove, a screened porch, and an outhouse—modest but sufficient.

What were they thinking? Why were these houses not built on piers as well? Lack of interest? Ignorance? Lack of finances? No, I suspect it was faith; faith that government levees would never fail. But was it false security?

Houses along most of the shoreline of Reelfoot during the 1912 levee breach were flooded nearly to the ceilings of buildings. High water, if not an honest to goodness flood, could be expected about one in ten years until the mainline levee of 1917 from Tiptonville to Hickman was built high enough and strong enough to withstand the latest record crest of the powerful river. It was false security.

Levees were built between the Mississippi River and the foothills of Chickasaw Bluff, which protected much of the floodplain farms and lake dwellers. Before the first government levees and the Lake County Levee District, floodplain farmland had to depend on private levees.

Every last one failed.

Regardless of the Mississippi River Levee, the lake itself might rise three feet or more some years above normal summer pool. But we were in for some dry years during the 40s through the 50s. The river might not flood at all during dry years. So, the reason neighbors and kin folks built their houses flat on the ground might not have been entirely from faith in government levees—it could also have been a simple lack of memory.

Conquering nature, Manifest Destiny—the country never lost its zeal to tame the western wilderness. Much of that ambition and energy seemed to settle along the Mississippi River. Since the 1927 flood, the river channel and levees have been remodeled and rebuilt several times—deeper or higher, wider and better armored each time, examples that demonstrate the power of hubris and determination over nature, each step designed to defy the threat of a new river crest. Many politicians were elected because levees were high on their list of promises, although at Reelfoot few anticipated the lake getting into their houses. So whether the levees held or not, promises to build bigger and better ones were convincing enough to get votes. Besides, for some, the good thing about levees was that they provided work, especially "slab fields" to fabricate concrete slabs. Concrete slabs armored the levees, and the plan was to build them even higher and stronger than any before. So, it is true, levee construction did hold back some floods and provided a lot of jobs during lean years.

The downside of levees was never mentioned, unless they failed.

But the levee breach of 1912 was soon gone and forgotten. Levees have held now for nearly eighty years. What was left to worry about? If there was any concern, it was hidden; the way to avoid worry for the farmers was to plow another field; for the Lake People, bait trotlines, knit, or set a few nets. Tending one's own business *seemed* the answer to a lot of problems.

Although floods during my young years were not as severe as those of 1912, there were a few reminders. Stepping out of bed into ankle-deep water was one. But Mom's cheerful attitude kept away low spirits; cooking breakfast while standing in ankle-deep water was a novelty for the young. The spirit to cope with it was contagious; the entire flood inconvenience seemed to create more neighborly spirit than stress. At least, that's the way I remember it.

Once the house was secure enough to carry on and comfort the youngest children, young men and their fathers were out in boats with fishing tackle or helping neighbors. Some of the family might move in for a few days with friends and relatives who lived in the hills. We never moved. As I have said, floods just brought out Mom's tough streak. She'd rally her brood as soon as the water fell. With more cheer than gloom, the mess would simply be cleaned up, disinfected, new wall paper installed. Within a week, the house was restored. Then, we'd carry on as if the flood never happened.

Old timers knew very well that flooding from the river regenerated the lake with fish and good health—but not all understood how much it contributed to their livelihood. An abundance of fish meant more food for furbearers. Food sources low on the food chain, such as mayflies, drag-onflies, and chironomids (woolies), were substantially increased, which meant more food, food for bullfrogs, turtles, snakes, birds, and other predators higher up the food chain—and ultimately food on our table.

That's the way it is; floods are essential for normal floodplain ecosys-tems, and do more good than harm. That's the way ecosystems work when given a chance. A problem comes when it is disrupted unnaturally; say, to prevent the natural rise or fall of lake levels, or to tolerate the invasion of exotic plants and animals. The careless introduction of exotic Asian carp and water willow that have invaded Reelfoot, for example, are believed to have compromised native plant and animal communities. These fish are suspected in the disruption of sport fish populations as well in food com-petition and spawning habitats. And there is little doubt that the invasion of water willow has compromised nesting and brood-rearing habitats for birds, and breeding, feeding, and shelter habitats for furbearers. Un-known is the ripple effect these invasions have on the total Reelfoot Lake ecosystem. Tennessee Wildlife Resource Agency (TWRA) biologists and managers have taken steps, however, and hope to resolve these issues.

When you live in a swamp, you expect it to be wet. So you adapt. Some families built boardwalks here and there to avoid wet feet. Half a dozen boardwalk bridges were built across the bayou to their houses. Mr. McQueen (Brothers Harry and Grover) provided a boardwalk across the bayou from their store, and others were built at strategic locations along the bayou. There really were no private boardwalks because no one who used them was concerned.

Beyond our backyards was a wonderful wilderness: a quarter-mile band of hardwoods, a strip of willows, a marsh dominated by giant cut-grass, and then, the open water of Upper Blue Basin. We hunted, fished, and trapped out of our back doors. Nearly every head of household came

Figure 17. Tennessee Academy of Science biological station.

from a generation of commercial fishermen, trappers, hunting and fishing guides, buyers, or all of the above. Very few sought to leave that life, and those that left daydreamed about it.

Walnut Log Road, on the east bank of the bayou, beyond the lodge, continued south another quarter-mile as a semi-gravel/gumbo dirt (or mud) road. The main road ended at a curious looking building owned by the Tennessee Academy of Science; to us, it was "The Bug House." Here again, the building was on eight-foot cypress piers. It was covered with the same material and paint as the Walnut Log Lodge. University professors and students often lodged here to conduct studies or collect specimens on the lake. Peeking through the windows, one could see a collection of formaldehyde jars holding specimens of frogs, snakes, fish, insects, and the Lord only knows what else. What went on here generated a lot of curiosity and eventually became of great interest to me. I sometimes provided the researchers with specimens, or took them on trips for biological collections. Somehow, most of it seemed to be in harmony with nature.

Just as intriguing was a boardwalk that came out of the second story porch of the Bug House. It crossed the Bayou du Chein, and went some four-hundred yards west through the woods and marsh to Upper Blue. The ridge the boardwalk landed on is called Goat Island. It goes south from the Walnut Log Canal to Mud Basin. The February 1920 issue of *National Geographic* reports that this island was owned by the Ranger Oil

Company. The big news was that this area, like Caddo Lake, Louisiana, was rich in oil reservoirs. Much of the island was sold off to investors in ten-by-ten foot plots. The rush to the oil fields finally ended when it turned out that the celebration was based on faulty information. Presently, the island is in state ownership. Imagine living in a place with such adventure. A building full of nature specimens and a boardwalk—the nearest thing to Heaven for young explorers. It certainly aroused the curiosity of the bayou kids. A significant opportunity was missed in not building several more of these boardwalks around or across the lake.

President Franklin D. Roosevelt's New Deal Civilian Conservation Corp (CCC) Camps during the late 1930s and early 40s was a grand idea for our great outdoors. They built a lot of park facilities, including boardwalks, some of which still exist at Blue Bank. One is the Round House. My father helped build it during the CCC days. Of course, building and maintenance for meager park budgets is difficult, but it's something of our nostalgic past that provides quality outdoor recreation and education. In fact, the rising interest in leisure outdoor recreation and conservation education almost seems to demand a revision of facilities like trails, boardwalks, and day-use facilities. I know of no other lake, then or today, more ideally suited as a boardwalk premier in the Southeast than Reelfoot Lake.

The road to the Bug House and beyond was nothing more than an ancient logging road, where the ruts of loaded logging wagons seemed still visible in the gumbo dirt. (Gumbo, you need to know, is a clay and alluvial river floodplain soil with a consistency between Silly Putty and glue—tough on hikers and vehicles when wet but rich in plant nutrients). The road entered the US Fish and Wildlife Refuge Grassy Island Unit at

Figure 18. Keystone loggers: A mule team load of logs.

the Bug House and went through a "sure enough" wilderness of cypress swamps and hardwoods for the next two miles. It ended at Brewer's Bar, the original site of the old Blue Wing Hunting Club. It is a paved road today and a splendid place to enjoy a hike or slow auto tour. Woodlands along either side of the road (except an Indian mound and two small fields) are more cypress swamp than dry land. I could imagine a team of mules straining on a wagon load of huge logs fresh out of the swamps. That was a common sight in the early 1900s. Bobcats might still be seen on the road, great blue herons as they flush from their crawfish hunts. Flashes of white are common, as whitetail deer flags bounce off into the trees, and yellow warblers flit here and there through the trees. Two neat wood trails with boardwalks along the road today supplement a trip here and provide a half-day outing.

Growing up on the Bayou

For a kid, there were more things to do at Reelfoot Lake than time to do them. Fishing, of course, was one of those activities. Without a guide or much advice from anyone, it seemed natural for me to rig a fishing pole and fish when the Bayou du Chein was less than twenty yards from our front porch. My earliest memory is of sitting on the bank of the bayou with a fishing pole in my hand.

Fishing on the bayou filled many wonderful days during my early youth; and while it was an adventure, it was also a source of the family's favorite meal—fried fish. This was especially the case during summer since commercial fishing was mainly during the colder weather of fall through early spring, except catfishing. The dads (and sometimes, the Lakies) guided and had other duties during summer. A special day was when Dad got a break and took me bluegill fishing out on Upper Blue. But mostly, it was me, the bayou, and the fish.

Ever since I was somewhere between the age of four and maybe eight, Mom had a problem keeping up with me. My wandering often took me beyond her beckoning call. She had a sixth sense—an aura of some keen awareness—that seemed to tell her when I needed to be accounted for. It usually ended without much consequence; a stringer of fish as witness to my good intentions generally settled her concern, and I survived to fish another day. It always helped that the family loved fried fish; it was the size and variety of my catch that sometimes raised questions.

Otherwise, off I'd go on a typical day with a fisherman's usual enthusiasm. Always ready was a rigged switch-cane fishing pole at the corner of our front porch, the first thing I'd likely pick up on the way out. The next thing was to dig a few worms. Not so easy during dry season of late summer. Worms or no worms, there were alternatives, and the fishing trip was rarely cancelled. The bayou had plenty of bait—if I could catch it. Sometimes it took much of the day to poke around in the aquatics prospecting for critters that might be bait. One was crawfish. Tin cans hold an amazing fascination for crawfish; they simply adore tin cans for seclusion, upscaled aquatic houses.

Along the bayou, I might stop and talk to Mr. Verge Powell as he sat whittling a paddle with a saw blade knife from an ash plank. He'd hang it on the wall of his shanty when finished, along with willow push-poles and another half dozen paddles he'd carved. "Got to whittle down-grain, boy, if you want to do it right," he'd say. That might be the only thing said.

Then, I'd mosey on up the footpath along the bayou beneath the tall cypress trees. It could take half of the day to round up enough cans for the crawfish traps, set them out, and raise them pretty soon afterwards. Usually, I could expect one per can. One crawfish tail might make three or four baits. Those not used went—well, need I say it; they sizzled with the fried fish I planned to catch, only if Mom allowed. I loved warm, rainy days. Fish also liked them. Rain flushes insects, their larvae, and worms into the water. Fish are cued in the first sound of thunder. Barefoot with no raincoat and usually no hat, I might be fishing on the banks of the bayou before the first raindrop.

Strange that a crawfish cuisine, so popular with Cajuns, only raised eyebrows from most of the populace around Reelfoot. I mentioned it to Mr. George Scheland, a skilled commercial fisherman who lived on my side of the bayou. Mr. George lost a leg a long time ago for some reason never mentioned. Getting around in a stump-jumper and tending nets might seem a problem for him, but he managed remarkably well with one crutch. He stopped knitting on his net long enough to stare a hole through me. "You'd have to find me starving to death, boy, to catch me eatin' a danged crawdad!" Then, he went back to knitting. Finicky eaters about some things, you'd think my good neighbors were offered snake meat. But one might get a fork in the back of their hand reaching for BBQ possum or ground hog with sweet potatoes. I've known Lakers to eat owls, great blue herons, and starlings—but crawfish and snakes? Never mentioned.

Why not crawfish? No idea. But that taboo wasn't adopted by the small adventurous band of "swamp rats," of which I was a founding member.

We often spent nights out on the lake as barely teenagers, just for pure adventure. Supper could be anything available—turtles, frogs, mushrooms, poke salad, snakes, persimmons, and so on, not to exclude crawfish. I once tried dragonflies. Just once.

But the fish I caught were often half the size of a sanctioned keeper—bony little fish. Fried crisp and crunchy—very tasty. I kept them all. A fish large enough to take my hard-earned bait meant it was large enough to be skillet-fried. And I nearly always had a fairly good stringer by the end of the day. But Mom was not always up to frying a skillet of small, bony fish in the middle of the afternoon. "Son, don't you think those fish should grow up a little?" Yeah, some catches were pretty small. But, if Mom detected so much as a disappointing frown, she'd likely smile and hand me the fish scaler and a pan. "You should take them to the outside to the fish-cleaning table," she'd say.

Little bones were the problem. For some folks, little bones could disrupt a fine fish dinner, but hardly a brief moment for us. The troublesome ones seemed to hang out around tonsils, thus, usually visible and easy to reach with the right tweezers. Mom had coached us as youngsters about the protocol for eating boney fish. We figured out the technique in no time, and there was little reason to worry—Mom was a skilled emergency fishbone surgeon. She kept a flashlight and forceps handy in the kitchen where the instruments were stored in a Maxwell House Coffee canister—her medical kit. The surgical extraction of a tiny fish bone lodged in a throat was one of her specialties. In her absence, however, a hefty gulp of bread might do the trick. Apparently, most fish bones can be digested quite easily. Any concern with fish bones beyond this, I don't remember. I never knew of a doctor invited to extract a fish bone—and rarely for any other medical needs. Doctor visits were usually during the bluegill spawning season of May and June when the sole purpose was to extract a stringer of "gills" from hollow stumps.

There was this other need—someone to cook bony little fish when Mom wasn't up to the task. Miss Lula Scheland, our beloved neighbor, was the answer to that. She lived with her husband, Henry, only a short distance up the west bank of the bayou, the same side where we lived. It did not matter what time of day it was, or the kind or size of fish I brought—when I went to her house with my catch. Miss Lula happily obliged. She'd help clean fish, meal, pepper, salt, and ready them for the skillet. They soon sizzled in her big cast iron skillet, blackened and seasoned from many fish fries. The aroma of fresh fish or coot and gravy seemed made for bayous; the unmistaken smell of these cuisines follows the bayou for a mile up- or downstream. So, the entire neighborhood knew when Miss Lula was cooking. But it was rare to find someone cooking in the middle of the afternoon. If so, it was probably Miss Lula cooking fish for "Ras" Johnson's boy. You can call them skillet-fried, sautéed, or pig-grease-cooked; either way, fried fish got no better than Miss Lula's. With a slice or two of Wonder Bread, a green onion, and a cold glass of sweet Lipton tea, we usually sat alone and ate until the platter was clean.

Too soon, I started to school and had to adapt to a lost life of leisure. Long days of summer fishing faded away like the setting of an August sun. With school, daylight-free time took a hit. Nevertheless, the call of the wild in my bones ignored it. Darkness rapidly approached by the time the school bus dropped us off back home. Fall advanced rapidly, and trapping season began. So in the dark, I raised a mile of trap lines. Back at the house, I pelted and stretched the catch on boards before bedtime. I caught thirty mink one season. Each pelt was worth twenty to thirty dollars, enough to buy nearly all my school supplies and clothes for the year. Self-support was a good thing because our family had grown, and Dad had two more of my sisters to think about.

Winter hunting season came and went. As school, trap lines, and a little duck hunting were fast-paced, winter was a whirlwind of excitement and happy times. As spring warmed the lake, hungry snapping turtles emerged from their winter hideouts and were on the prowl, and there were turtle nets to tend to. These had to be raised before the early morning school bus ran. The goal was to fill two or three cypress-boarded liveboxes half submerged at the boat dock with twenty-five or so snappers, and clean them the following weekend. Some buyers wanted live turtles, some wanted them dressed. Taking apart a snapping turtle is like a jigsaw puzzle in reverse; taking one apart in serving-size portions is a slow, meticulous job. It begins with disabling the dangerous jaws and removing the tough shell. It gets a little easier after that, but not much. Some say there are about ten different kinds of meat in a turtle; I can find only half a dozen. Sometimes it took most of a Saturday to clean and ice-down the catch from last week.

So, you'll quickly surmise that summertime to a Gumbooter kid was something special; you'd be right. Many days were spent on the lake catching turtles of some sort, or attending trotlines for catfish. Local buyers would pay five or ten cents each for baby turtles; I might catch fifty or a hundred in a single day. I was told they went to pet shops, zoos, or colleges and universities. There were local buyers, but Dad sometimes shipped them to our own markets. Catfish trotlines kept me busy from late summer through fall. For a kid, a half-mile line with three hundred hooks was about all one could handle. There were simply more things to do during summer than time allowed—and only some of it was done during daylight hours.

Boys and Frogs

Reelfoot has little frogs, medium frogs, and bullfrogs. As a kid, there was maybe a half dozen "froggers" around the lake. At one time or another, we collected all of these frogs, some for play, some for fried frog legs, and some to sell. It was not unusual for Dad to wake up some warm night in May because his hearing was tuned to the roar of bullfrogs, maybe a mile away. He might say, "Why don't you go get us a dozen or so of those frogs for breakfast." Or, he would say, "Let's go get a few of those big boys before the spawn is over."

As kids learning the ways of our fathers, we collected the little ones (the cricket frogs and spring creepers), pretending we were on a major bullfrog hunt. We even dressed the little fellows for a make-believe frog meal. That pastime ended when one of my best friends, Spencer Powell, on the spur of the moment decided not to sacrifice a particularly colorful specimen. But my rusted, double-bitted ax with a short broken handle was already committed. It left the tip of his right hand forefinger dangling. Spencer did not even whimper; he slowly reconnected the finger in the grip of his unwounded hand and walked home in silence.

A few days later, Spencer came back to my house with a big smile. The finger was tightly wrapped and the last digit appeared intact, somewhat the same as before. His daddy had doused the finger with kerosene, splinted it, and wrapped it up. That ended our make-believe frog hunts but added another salient memory neither of us will forget. (Today his finger is fine, but with an odd crook at the last digit.)

May is peak spawning month for bullfrogs. A kid might do as well as a seasoned hunter and could make fifteen or twenty dollars some nights gigging frogs—the adventure, of course, was worth far more than that. Many times, I, and sometimes a second mate, would be parked at sundown at the outlet of Walnut Log Canal on Upper Blue Basin. The objective was to fill our live-box with eight or ten dozen bullfrogs; it could take all night.

Waiting with headlights, live-box, push-pole, a jug of water, and a cold jam or baloney and biscuit sandwich, we'd listen intently to get a bearing on every group of spawning frogs within hearing; on a still night, that was easy to do. Sound across still water, especially the crescendo of spawning bullfrogs, can be heard a mile or more, which was about the distance from the canal to Long Point or Brewer's Bar. One might hear two or three large groups of fifty adult frogs or so on a good night.

Once a group of "spawners" were selected, we'd make all haste to get there, picking a second choice on the way, just in case things didn't go well with the first. Hopefully, spawning frogs would be in lotus pads. If they were in the tall spatter dock or button bush, they would be difficult to see or approach because the bushes were too high or too dense to catch frogs. These summer nights were usually so calm (at least you hoped), one could hear the bow of the boat slicing water over the "click-clack" rhythm of busy oars powering along the boat. Tonight, we headed south toward Brewer's Bar. I could smell fish frying and the lonesome sound of a tin pan clinking somewhere over by Gray's Camp. A lone fisherman, I'd gather, had made his catch today and was preparing supper.

If you have not been in the middle of a marsh wilderness with fifty or so bellowing bull frogs, you've got to try it. You can begin the magic with a "click" of the headlight; deep darkness and stillness suddenly drape over the entire marsh, dark as a coal mine until your eyes adapt. Then the senses kick in; every movement, click, chirp, stir, splash, and flick of light is magnified. The first thing to notice is the unexpected brightness of the moon and stars—and then the fragrance of the swamp: Only a natural swamp has the musty aroma of sweet lotus blooms and their yankapin fruit, the fishy smell of frog spawn, and the fresh scent of chlorophyll rising from lush, green aquatic plants as they rub against the boat. Then, the little sounds—a chirping chorus of cricket frogs, a soft splash of something here and there. "What was that?", you'd mutter softly to no one. You sit quietly, another ten minutes, hardly breathing, soaking up every stimulus from song, curious sounds, and smell. Then the bigger sounds— swoosh! Something heavy parted the southern smartweeds that brushed the boat—a beaver? A forty-pound catfish? Could've been a large buffalo fish. But then, you'd have sworn a big bull frog just skipped noisily across the bow. Maybe not; an hour later, you begin to wonder; could it have been my imagination?

Likely as not, the urge to reach for the "on" button of the headlight is hard to resist: "Good Lord, who was that?!" The inquiry is silent this time.

But hold on; we don't have alligators; it was probably only a family of ot-ters, a muskrat, a beaver, or a snapping turtle. You never know for sure what creatures might emerge from a dank, teeming marsh of life and detritus during the black of midnight.

Suddenly, as if the conductor of the night raised his wand, the choir of many creatures rests; there's a brief but deafening silence—twelve beats to the count, the silence is instantly broken by the ubiquitous, high-pitched buzz of a zillion new voices, voices from the littlest of the unknown creatures—tiny frogs, insects, and significant others, strange to all but residents of the marsh. The crescendo ends in a quiet pause. Si-lence; the count is ten. Then, the lead bullfrog croaks one short verse. It's a cue; a new chorus of bullfrog base harmony begins. Headlight on: the hunt begins.

We used three types of devices to capture bullfrogs: metal prong gigs, metal live catchers, and bare hands. Gigs spear and kill the frog; catchers grasp and hold the frog alive; bare hands leave the option of bonking the frog or storing it alive. Most of the time, we wanted to keep or catch alive so they would not spoil before we cleaned them. It took a certain kind of finesse to slip upon a bright-eyed bullfrog sprawled on a lily pad, bloated twice its size. The captured air keeps the frog afloat, and helps it master its best rumbling croak. The would-be prey looks wisely straight into your light, as if it suspects you are from Mars. One wrong move and . . . "Poo-tush!" a deflating blast of bullfrog air, and the prize is gone!

Of the three, the most exciting is to catch a frog with bare hands. To get in range requires a "cool-hand-Luke" kind of skill. One person must push the boat strategically in the right direction with a light, and a strong, twelve-foot push-pole, suited for the purpose—and don't bump the boat! The catcher lies on the bow, headlight on, hands low over the water and vegetation. If your headlight is not too bright, and all goes well, one can be scooted through a bunch of frogs while grabbing them around the small of the back, one after the other, sometimes with one in each hand. It takes some calculated scurrying to turn around off the bow of the boat and get them in the live-box without an escape.

Landing the grab jaws of the catcher exactly in the small of the frog's back requires steady hands, and that comes with practice. What can set your steely nerves on edge is the competition—snakes out to catch supper. Sometimes it's competition for the same frog—at the same time! Not to worry; there's only one snake on the entire lake that deserves concern—the sassy cottonmouth. Even the cottonmouth is of little concern if it has

Figure 19. Barefoot Boy and his dog
on Bayou du Chein at Walnut Log.

already captured a frog; during that time, all snakes seem oblivious to the
rest of the world as they are singularly preoccupied with supper.

Bull frogs are not easy for a snake to swallow, especially during spawning when it blows up like a balloon. The reptile's unhinged jaws hardly
seem adequate for the job, but given enough time, it manages to get it
done. All of this is good advice but easy to forget when stretched over the
bow of the boat and hovering only inches above a light-blinded, bewil-
dered cottonmouth.

But the adventurous night on the lake must end by daybreak. It is usually a long way back to the boat dock and the sun has risen above the trees.
You could be satisfied that the night's work has produced a box full of
frogs. You would rather no one was at the landing when you arrive. But it
is early in the morning, and no one is likely to notice that your clothes are
slim, and you smell a lot like a bullfrog. A pleasant weariness has set in
and about all you can think about is a cup of hot coffee and sleep. But the
lowest part of the adventure is yet to be done; you must clean, package,
and ice-down maybe ten dozen bullfrogs before anyone welcomes you into
the house. By then, the warm morning sun is often closer to noon, determined to put you to sleep the moment you pause to rest. A bullfrog hunt
can be an intense enterprise.

It certainly was a great life for a kid.

Snapping Turtles Don't Bite Off Fingers

Daylight had not quite broken when I pushed the boat off of the landing. This was the year I set out half a dozen baited turtle nets to be raised before going to school. We had moved from the west side of the bayou to the east side, where the gravel road ran. The house sat on a high spot of ground ten yards from the bayou. Built about 1900, it the largest house in Walnut Log, once owned by Mrs. Myrtle Powell. She and Mr. Powell had passed away some years ago. Ms. Myrtle was a very cordial person and enjoyed conversation. She very well remembered the fateful night of October 1908; it was the Powells' house to which the Night Riders paid an unwelcome visit. I don't remember her speaking about those troubles. The Powells were the only family with the insight not only to build their house on piers, but also on ground higher than any nearby. So it was a lovely place to live, and a nice snapping turtle swamp was but a short quarter-mile distance west beyond the bayou—not so far that I could not raise my turtle nets and return to the landing in time to catch the school bus with my brother and sisters at the junction of Walnut Log and Bennett Johnson Road.

Crossing the Bayou du Chein and moving into the swamp, I soon arrived at the first baited turtle net; it was loaded with three big snappers. These nets were always set with part of the net out of the water so turtles could get air. I saw a few big snapper heads poking out of the net some distances before I arrived. If you would like to know how the snapping turtle

Figure 20. A baby alligator snapping turtle.

got its name, just stick around when it gets agitated. My nets were unusually full of snappers, and I ended up that trip with about a dozen nice
turtles. So it took more time than usual to get the rowdy turtles—clawing
and snapping, hating humans worse than a half-starved hyena, just for
being there—out of the nets and back at the dock, where I held turtles in
large live-boxes. I was on my way with a dozen turtles dumped loosely in
the bottom of my Reelfoot stump-jumper.

Time out!

A wise turtle trapper doesn't ordinarily do this. You usually put them
in a tow sack and tie the top to avoid the treat. But, I had not a single sacks.
The only practical way to corral these turtles was to keep them kicked
back while rowing the boat. Moving at top speed, I heard the school bus
come off the Chickasaw Bluff, half a mile away, toward our house. Routine
was for it to go down the dead-end road by our house, pick up more kids,
then return and pick us up on the way out.

On the way back to the dock and landing, the bus passed our house.
With luck, I would make a hard landing on the go, dump the turtles one-
by-one in the live-box, run like a streak to the house, get my books, and
make it to the pick-up place at the junction of the roads. Watching and
kicking back angry turtles that weighted ten to fifteen pounds while oar-
ing a boat can be tricky business. As I watched, the largest turtle crawled
up into the dark bow of the boat, leaving its tail toward me. That was good;
turtles get quiet and secure in tight, dark places. Their tail is a handle.
Only a grinning nut would try picking up these primitive warriors any
other way.

I heard the bus turn around about a mile away at the dead-end road,
just about the same time the bow of my boat touched our landing. The
panic was on! I had about three minutes before the bus would be back.
Grabbing angry turtles by the tail as fast as possible, I threw each into
the live-box. I was no longer worried about the big turtle; it had settled
down under the dark hood of the bow; that is, the last time I looked. But,
lo—what could happen but "Peter's Principle": instead of staying put, the
cantankerous reptile had moved, and I was oblivious. In one of my weak-
est moments, the lowly demon of the deep had turned around and was
fully prepared to go to war! Confident in my first assessment of the chaos,
I reached without much thought under the bow of the boat for the turtle's
tail. Big mistake. Open jaws waited.

A fifteen-pound snapping turtle on your hand is a chilling sensation.
That's bad enough, but worse is the predicament you are in; think of a

Figure 21. Bennett Johnson cautiously puts a
snapping turtle in a fish live box.

vise with cold, sharp, and bony jaws. I suppose I was fortunate because I
could have put my entire fist in its huge mouth. But there I was, stuck with
my middle finger hidden by a large, bony mouth. I'd been told a snapping
turtle could bite off a finger: "Snap," like that, a finger was gone. At that
very moment, I would have bet a dollar they were right. But, if that were
true, why was I still unable to pull my hand free from under the bow of the
boat?

As it were, I was going no place, afraid to breathe. A cold, metallic sensation, like a double-spring steel trap, clamped tighter and tighter on my
hand with the slightest movement. Actually, it was not like a steel trap—it
was worse! I've had steel traps on my hand. So there I was, stuck, no help,
and no way out. Upon reflection, it reminded me of a pup with a crawfish
on its nose. I could do little but look pitiful and be still.

Suddenly the role had reversed: the turtle had trapped the trapper. Dad's true character was not to get excited about yelling around our place. Anyone else would have thought I was being mauled by a panther. Not Dad. Rather casually, from the top of the small mound where the house stood, he asked calmly what the problem might be. The school bus had already left our stop when Mom threatened him with a broom to go to my rescue. That was the good news.

It took two trips and two sets of pliers before Dad broke the turtle's jaw and my hand set free. The school bus was probably all the way back at Dixie School before I was free. It would have been the highlight of the day had the bus waited long enough to pick me up and the other kids had heard the story. So instead of going to school, I went fishing, a little wiser about snapping turtles. All considered, it turned out to be a pretty good day.

More Adventure on the Bayou du Chein

The Bayou du Chein is an odd adventure on its own. It runs through all four major basins of the lake but originates near Hickman, Kentucky, where it is beheaded by the Mississippi River. The lower part flows south through Tennessee and on through Walnut Log. Kentucky folks call it Running Slough in their state, but the upper part they call the *Bayou du Chein*. It is a wonderful stream for canoeing, birdwatching, photography, and about any other outdoor pleasure. One place to begin is at the Long Point Refuge boundary road in Kentucky and float downstream. The bridge just east of the Long Point Refuge entrance is the Bayou du Chein. A drop-off here will make a great canoe trip through Walnut Log, Grassy Island, Gray's Camp, and beyond, if desired.

The bayou parallels Walnut Log Road and follows it south a couple of miles through the wilderness of Grassy Island, but it turns sharply west a short distance beyond Mud Basin and crosses the lake west toward Gray's Camp. Old timers have said that huge log drifts piled up across this end of Upper Blue Basin when the Mississippi floodwater ran through here before the levees. These massive drifts diverted the bayou from the east side of the lake across to Gray's Camp on the west side. From here, it goes south for six or seven miles along Choctaw Woods and Burnt Woods, and through Buck Basin. The channel joins the old hidden Reelfoot River channel here before it passes through Joe Basin near the shores of Samburg, and on to its outlet at the Spillway. From the outlet, the stream today becomes Running Reelfoot River, a channelized ditch that drains marginal cropland (a former fish brooding grounds, known as "The Scatters") before it goes south to the Obion River.

At Walnut Log, the lively little river was seven to eight feet deep and forty feet wide when I was a kid. When spring rains came, it had a feisty current after any rain more than about 1.5 inches. Back then, other than a commercial-fishing skiff, our Reelfoot stump-jumper was the only boat we had for work or play. We really didn't think another boat was needed since it was designed especially for the hazards of Reelfoot Lake—going over stumps and logs, and through willow thickets and fields of aquatic vegetation. The boat was built tough with heartwood cypress boards and a sharp nose at both ends, and bow-facing oars. It could probably carry a half-ton load and do fine with a push-pole and seasoned operator. But the stump-jumper required extra effort with the oars to power upstream.

Bayou du Chein can become a torrent during heavy runoff in late winter and early spring. It is a time when a commercial fisherman on the ball has his set-nets ready. That's when it seemed every fish in the lake wanted to go to Kentucky. I didn't know for sure, until years later, why they did this, especially during late winter and early spring. The clue was female fish; they were fit to burst and ready to spawn they were so full of gravid eggs.

That, of course was a clue to the mystery. Strangely enough, fish did not intend to stay upstream because they migrate; they went mostly upstream during spring and mostly downstream at other times. It took a while before I learned that fish going upstream during spring were headed to small lakes and sloughs to protected spawning grounds; those going downstream were headed for the lake or the river. It was a bit complicated, but it made sense that mature fish needed the security of shallow wetlands with plenty of cover to protect their newly hatched fry. Once their small fry grew large enough to survive on their own, most would leave the brood grounds and return to the larger lakes and rivers. That way, both the lake and the river were replenished. That's how rivers and lakes complement each other: rivers provide brood fish, lakes and marshes provide spawning and nursery ground; young fish grow to migrate and restock the river (and the lake), and the cycle continues.

I knew very well how tricky the current of the Bayou du Chein could be. But one cold, windy day in late February 1950, I went with my dad to raise set-nets half a mile up the bayou. This was a time when all kinds of fish went upstream to feed and spawn. Crappie, bluegill, catfish, and fat buffalo fish were the favorites, and Dad had a hankering for buffalo ribs, which have no small bones to contend with, and the meat is sweet and robust in flavor. Fried nice and crisp, no other fish can take its place.

It had rained "cats-and-dogs," a good four inches the day before, and the bayou was as rowdy as I'd seen, a muddy, swirling turbulence, really on the move. Dad worked at the oars of our stump-jumper, skillfully maneuvering to make every stroke count in order to buck the current with the least amount of energy. Still, a short break for a little rest was on his mind when we reached Dehart's Store, only a short distance upstream. It sat conveniently located between the gravel road and the bayou for either automobile or boat. The one-room store with a small storage room out to the edge of the bayou was small but adequate to supply basics needed by the community. Most important to me was the glass globe peanut machine. A couple of pennies and two turns of the knob could get you a good handful of red-skin delectable. My pitiful stare-at-the-peanuts look was rewarded when Dad fished two pennies from his pocket. Both of us being refreshed, we reloaded in the boat and continued upstream to the nets.

The only seat in the boat was the one Dad was sitting in; any other would just be a stumbling block trying to tend the nets. Besides, it would take up room needed for our catch. As habit would have it, I sat on the bow of the boat, usually permitted only to seasoned Lakers. We had gone only a short distance before the boat hit a heavy log, carried by the swift current and hidden dangerously beneath the muddy water.

Bam!

And off I went into the icy, mud-churning water. Dad rushed to the front of the boat to rescue me but the boat immediately spun out of control in the current, and downstream it went wildly drifting without a pilot. As Dad told it, I was nowhere to be seen. Looking desperately for signs of life, he noticed the ten-foot bowline was overboard. Inexplicably, he noticed the rope jerked a couple of times. Only desperation and reflex made him reach for the line. Miracles don't happen every day. But when he hauled the line in, my boot was wrapped tightly at the end—and my foot was in it! Somehow the bowline, despite all odds, had become entangled around my ankle when I went overboard. (Oddly enough, it happened at the exact spot where Dad had retrieved a neighbor who fell from his boat a few years back when the bayou was peaceful. I only knew him as "Mr. Frog." He drowned here in eight feet of water for unknown reasons.)

I'll swear to this day, in spite of the muddy water, my nose only inches below the bottom of the boat, that I could clearly see the lap of the tin that armored the boat, and small nails spaced neatly along the seams. Funny how little things like that imprint in one's mind during a crisis.

Wet and shivering, and still with a handful of soggy peanuts—no way was I about to discard them—I finished eating well-washed, half-salty nuts while Dad hustled us upstream a quarter of a mile to my Aunt Ivy Henson's house on the west bank of the bayou. There, a warm potbelly stove and dry clothes belonging to my cousins waited. About an hour later, we continued up the bayou to raise the nets; they were absolutely full of fish. It took both of us to haul the nets into the boat and shake out the fish. When we emptied the last one, the boat was filled nearly to the rails. I was afraid that even if a winter sycamore leaf landed on the boat we might sink.

Gumbooter Turns Farmer

Soon after Dad returned from the war, the government offered programs to help jump-start returning troops. Anticipating risky fish and fur markets, and a heads-up about other economic indicators, he, like many other discharged soldiers, attended programs that encouraged farming. It was probably a great idea to get soldiers back on their feet and to help calm the traumas of war. It did nothing, however, to brighten my spirits: to this day, I feel the sadness of moving from the lake to our first farm. Removed from the banks of my beloved lake, I was a duck with no pond, a hybridized "Son of the Swamps," half Gumbooter, half Kneebooter at the very best. Nice gardens, horses, goats, cows, and chickens were fine, but I had lost my legitimacy. Of course, that was stretching it a bit. After all, our farm was only a mile from Johnson Basin, an oxbow of the old Bayou du Chein River—a thirty-minute walk and a fifteen-minute boat ride through Walnut Log to the open lake. Still, I was uneasy; water and a boat had been the first things I saw every morning since I was born. Now, I could not see the water, let alone my boat. Besides that, I would probably be called a Kneebooter. There were adjustments to be made.

By and by, high school days were over, and my farming days passed. It also seemed to be a time when bread-winning occupations were going downhill. I thought of Alaska; my friend since birth, James Carter, had been reading magazines about the North Country; we might just hitch a ride and go there. By chance, that would be a good second choice. There, I could buy supplies, along with a rifle and good dog team, and head to some remote trapping grounds. Maybe I would find a soulmate, a lovely native maiden in the region, an amiable woods-woman with skills I might not know—one who didn't mind a little wilderness life, the smell of bear grease, and dogs in the house. It all went up in a puff of smoke; I joined the US Air Force. The daydream lasted for years, but my longing to be near the lake never left my mind.

Chapter 7

The Natural Resource Managers

USFWS Refuge

Recognizing the need for waterfowl refuges nationwide, Tennessee negotiated a seventy-five-year lease in 1941 with the US Fish and Wildlife Service. The refuge includes wetlands in the Grassy Island area and land in the Long Point area of Upper Blue Basin. Peak mid-winter count of mallards (an index species) has reached four-hundred thousand—but it's usually closer to half that number. Yet the lake is known for its global significance of all birds in the flyway. The lease between the state and the Reelfoot National Wildlife Refuge (RNWR) was renewed in 2016, with probably even more concern for other birds in the flyway. Today, the refuge has some 10,428 acres (the northern one-third of Reelfoot) within its boundaries, some of which is in Kentucky.

Reelfoot National Wildlife Refuge office is located near Walnut Log and Reelfoot Creek on Highway 157. The scenic setting with office facilities and a small museum overlooks Grassy Island just beyond the open land. The museum has very well-done exhibits of wildlife and items that reflect the setting and refuge mission. The friendly staff has frequent nature/ conservation-oriented programs. Out back is a short walking trail.

You'll be happy to find that all RNWR areas are open to the public, except special regulations and areas that are closed from November 15 to March 15. In addition to its attractive museum display in its headquarters near Walnut Log, the refuge has a wildlife watch tower in the Kentucky part of the Long Point Unit and two excellent boardwalk trails in the Grassy Island Unit. The Long Point Unit has a wildlife viewing tower open to the public year round, where thousands of waterfowl normally swarm and can be seen during peak waterfowl migrations.

The future of Reelfoot depends heavily upon its natural resource managers. Reelfoot National Wildlife Refuge (RNWR) is one of the agencies. Its partnership with the State of Tennessee was a wise one. Strategically located along the world's greatest flyway, the refuge's primary objective is to enhance and protect migrant bird populations. But there is a bonus: the refuge also protects native animals and plant life, and provides outdoor recreation and conservation education for people everywhere who value nature and the great outdoors. The objectives of the RNWR, TWRA, and Reelfoot Lake State Parks (RLSP) are complementary. All objectives of these agencies involve outdoor recreation. Compatible outdoor recreation and wildlife management mix very well; each depends on the others. Without support from all sides, none will succeed. That's the reason this cooperative arrangement can hardly be more helpful to the objectives for future natural resources, especially for Reelfoot wetlands. Regretfully, I must mention that the cooperative agreement was also an important stabilizing document against whimsical political pandering. Political support is gravely important to the future of natural resources and outdoor recreation, but can cause a dismal failure when it serves only a vested interest, and not the public good. The initiative and wisdom to make it work depends on good managers, an involved public, and political support.

Avoided for at least the last thirty-five years, a lingering and critical problem must be solved. Regrettably, also, is that a relatively small acreage of private wetland in Kentucky is in the Reelfoot floodplain—and there are no willing sellers. This means that major decisions for water management on Reelfoot Lake are at the mercy of a very few people. It is a political problem that can only be resolved by the initiative of the players: the landowners; Fish and Wildlife Service, Department of the Interior; the State of Kentucky, and the current US Senate chairman from Kentucky, who has never supported the future value or needs of Reelfoot Lake. Neither the landowners nor the Kentucky politicians involved show an interest in the management and preservation of the lake. Tennessee players have no jurisdiction. Ironically, the Kentucky Department of Conservation is more than willing to participate in the best management of Reelfoot. Kentucky politics are apparently ultra-sensitive to Western Kentucky landowners and their votes. Consequently, natural resource managers are at a stalemate on how to proceed with long-range management obligations.

Lake Isom is a satellite National Wildlife Refuge (NWR). It is located five miles south of the lake as part of the RNWR complex. The 1,850-acre unit was established in 1938. The entire area below Reelfoot—including Lake Isom—was one huge, shallow wetland system known as "The Scatters" before it was drained for agriculture. It was really a part of Reelfoot Lake, until Highway 21 (acting as both a road and a levee) separated them. The

Scatters was a tremendous wintering waterfowl area, a rearing ground for furbearers and birds, and a spawning ground for fish. It was like a living conduit of energy to Reelfoot Lake, where fish and other aquatic migrants could reach the lake from the Obion and Mississippi rivers. In fact, The Scatters was a year-round reservoir for all kinds of aquatic life. Nearly all of it was drained for cropland, a major reason to preserve Lake Isom.

The West Tennessee Wetland Corridor Project

The West Tennessee Wetland Corridor Project actually began as a cooperative waterfowl project between USFWS and TWRA to restore the waterfowl habitat along the Mississippi River. That project lost its momentum as its leaders diminished. After that, it was quietly considered a special, long-range project by a meager few (a vision TWRA director Gary Myers and I relentlessly hounded) to establish a wooded wetland corridor between Reelfoot Lake and Shelby Forest Park and Wildlife Management Area (WMA) near Memphis. The final version would provide a wetland corridor that would restore the historical habitat needs for migratory birds and wildlife native to the region.

Imagine the extent and super benefits of this project. It would effectively complement Reelfoot Lake projects and restore more than one-hundred thousand acres of lost wetlands in Tennessee along the Mississippi River. Today, the project begins with the northern boundaries of Reelfoot Lake National Wildlife Refuge, even into Kentucky, and continues southward to include Reelfoot Lake and Lake Isom NWR. From here, the vision continues southward as a diverse wetland to Shelby Forest Park and Wildlife Management Area near Memphis.

The leadership has changed since this unofficial project made headway, but there is reason to believe it continues to progress. All of these new lands are purchased from willing sellers. These are generally marginal farmlands, or cleared land frequently too wet to farm. Consequently, there should be few objections from the agriculture industry because the plan actually helps farmers already stressed by land too wet to be farmed profitably. TWRA wildlife management areas like Tumbleweed, Bogota, White Lake, E. Rice, Moss Island, and Anderson-Tully, downstream from Reelfoot, make up thousands of acres that now fit into this plan. The goal of the original purchases (some isolated by as many as three miles) is to eventually fill in the gaps between these areas to provide a continuous unit of mostly forests, fish, wildlife, and outdoor recreation wetlands from here to Memphis. That goal should not be diminished.

Consider the multitude of forest products: bottomland hardwoods, sustained yields of lumber, buffer zones to slow wind and floodwater, habitat for birds, fish, and other wildlife. Along with federal refuges, there would be thousands of acres of destroyed wetlands restored to functional and sustainable natural wetlands for fish, waterfowl, shorebirds, and water birds. The benefits will be enormous, unlike anything in Tennessee, or for several hundred miles north or south along the Mississippi.

The impetus for this project is long overdue: wetland biologists have estimated that, from the 1780s to the 1980s, the conterminous United States lost approximately 290,000 acres of wetlands per year. Half of these acres were lost during the beginning of global interest in the soybean industry, between the 1950s and 1960s. Nearly 90 percent of wetlands along the Mississippi have been converted to cropland and other uses, and about 85 percent of forested wetlands along West Tennessee rivers have been lost to agriculture and channelization. This project will go a long way toward state and federal wetland restoration objectives projected for the Lower Mississippi River wetland initiatives. The project will also complement Upper Mississippi River wetland projects. The Upper Mississippi River National Wildlife Refuge contains 240,000 acres of land stretched in parcels for some 261 miles along the Upper Mississippi River floodplain.

Someone had anticipated these changes, because this refuge from Rochester, Minnesota, to Davenport, Illinois, has been in place since 1924.

Wetland restoration efforts for this West Tennessee project have gone quietly unnoticed for about three decades. Although an ongoing effort, it is about time to raise it from the dust and present it to the public as an active project in need of expediting. One day it will be known as one of the most ambitious plans yet for the state's wetland recovery for lost natural resources and outdoor recreation potential in this part of the country, and Reelfoot Lake is at the head of this master plan.

What this project needs today is an expeditious boost from the new leadership and the public who will benefit the most—outdoorsmen, bird-watchers, hunters and fishermen, naturalists, boaters, eagle watchers, photographers, and all whose life is not complete without a generous dose of nature.

This project complements tourist industry objectives for the Great Rivers Road and the Mississippi River Corridor projects. Credit must go to TWRA commissioner Tom Hensley and the Tennessee Legislature for passing Tennessee's Wetland Acquisition Act of 1986. Thanks to TWRA director Gary Myers for supporting the recommendations of wildlife managers. The result has been the purchase of some sixty-thousand additional acres in the 1980s and 90s, mainly for waterfowl and shorebirds, and for reforestation.

TWRA Wildlife Management Area and Refuge

The Reelfoot WMA includes all public land not in state parks or in the National Wildlife Refuge. The WMA contains ten-thousand to twelve-thousand acres of open lake, cypress swamps, and marsh at normal pool, in addition to four thousand acres of bottomland hardwoods, including the Black Bayou Refuge. All of the WMA is open to the public, except for closures on Black Bayou Refuge during waterfowl season or special hunts.

The Black Bayou Refuge work base is in the unincorporated community of New Markham. TWRA purchased the land during the 1990s, primarily as part of a buffer zone around the lake to filter runoff and defray complaints about high lake levels. It also provided a good opportunity to meet waterfowl management needs—to improve waterfowl hunting around the lake. Providing multiple waterfowl flight corridors between refuges not only helps keep single-area concentrations of wintering waterfowl from exceeding carrying capacity (the maximum population the land can accommodate during a prescribed period), but also disperses hunting opportunities for a more equitable harvest of waterfowl.

The refuge is in two units, the New Markham and the Phillippy unit. The Phillippy unit is about a mile north of the New Markham unit. Until the 1990s, the refuge was about two thousand acres of bare cropland, mainly wetlands and marginal cropland planted with soybeans. As you pass through the area, compare it to the bare ground of adjacent large fields—it's no different than the former refuge. All of it was in forest when I was a kid. Since then, trees were cleared for cropland.

Black Bayou Refuge is something of a first model for the management of river wetlands. Waterfowl managers traditionally managed wetlands for waterfowl only. That philosophy has changed. Today, the policy is to diversify wetlands to provide as many collateral benefits for wetland wildlife as is reasonably possible and, at the same time, to meet the primary objectives of accommodating wintering waterfowl and improving harvests. In other words, the trend is to manage the entire wetland ecosystem for multiple benefits rather than of only its parts. Notice the wide vegetated corridors; they were planted in oaks, cypress, and other hard-woods between fields as travel lanes and food plots for upland wildlife. Deer, furbearers, wild turkeys, quail, squirrels, and other upland wildlife are now part of TWRA's wetland management program at no sacrifice to waterfowl management objectives. The trees today stand twenty feet high and produce a crop of nuts and fruits favorable to a variety of wildlife, including waterfowl. Low-level terraces instead of high levees are used to construct managed water pools for waterfowl, shorebirds, and wading birds. Today, at a distance, the area looks to be more like a forest than barren winter cropland—and that is largely the idea. An elevated wild-life watch tower about mid-way along the McCutchen Road is open year round, but the interior beyond the tower is closed during state waterfowl seasons.

Another reason to use the refuge as a buffer zone is to protect the integrity of the lake. Previous owners dammed off portions of the lower part of the New Mark unit for cropland before Black Bayou was in state ownership. Notice how the road cuts sharply northeasterly along the woodland (known locally as the "State Woods") and crosses the wetland. The young cypress trees on the left were planted about the year 2000. The road was used as a dam to prevent the lake from flooding cropland on the left side of the road. A large pump house and pump were used to keep lake water off of cropland. But that has changed—the pump has been removed and that part of the lake reclaimed. A large culvert now allows water from the lake to flow freely as the lake rises and falls.

Farming methods and programs have changed in river floodplains since about the 1960s when soybeans became a worldwide marketing product. Private farmland over the years has continued to clear land to the lakeshore—even below ordinary lake levels—for soybeans. Soybeans can often be successfully raised in wetlands where other crops cannot. That's because the ground dries long enough during late summer for short-season crops like soybeans to produce.

This creates conflicts.

Agriculture objectives and critical needs for managing lake levels are directly the opposite; farmers want the lake as low as possible during the growing season, fishermen want the lake as high as possible, and lake managers need multiple lake levels to protect and enhance the ecosystem. Since water levels management allows the lake to be maintained from half a foot to one foot above pool, some crops in these low-lying areas are at risk of flooding. Furthermore, sometimes nature exaggerates the en-

Figure 23. Lake freeze-up and mallards on ice at my dock.

tire scheme and causes the lake to rise even higher, usually to the detriment of the lake manager's reputation and purpose.

Actually, Black Bayou Refuge, as you have seen, was partly designed to change minds. Traditional methods for managing waterfowl in river floodplains are antiquated and destructive to rivers. So Black Bayou demonstrates in a small way how and why not to build large obstructions like high levees to impound water for waterfowl, or to protect cropland. But farmers, private hunting clubs, and wildlife agencies have used these destructive methods for half a century; of course, it was not their intention to destroy our rivers. There was just no one to consider the benefits of a natural river, how they functioned, or how to heal them when they were misused. There are good alternatives, however, and it's simple and more economical than trying to overpower or sacrifice the river: move incompatible developments above the annual floodplain, the main temporary storage and flow way for flood relief. Growing trees and allowing the river to recover is tremendously more valuable than net losses and destruction caused by misuse of the river. So, now, waterfowl managers have a better option: avoid levees and use low-level terraces instead; low terraces built on contours allow incoming streams to flow and floodwater to flow over the terraces unimpeded with little effect on the natural hydrology of river floodplains.

Observe the large expanse of open fields as you go from Black Bayou to the Phillippy unit of the refuge. You can easily see riding the roads of Lake County that the Mississippi River floodplain is a large expanse: look west, and notice the mainline levee; look east from Gray's Camp, and notice in the far distance the Chickasaw Bluff—this was the original floodplain. One can stand on the east levee and see the levee in Missouri across the river; this passage between the narrow confines of the two levees must carry the runoff floodwaters of today. The original floodplain without the levees was so wide one could hardly recognize the threat of flooding, because the water was less swift and much shallower.

One other thing: we cannot forget that rivers create wetlands; lousy river, lousy wetlands; healthy rivers, healthy wetlands. The health of the Mississippi River cannot be ignored because wetlands like Reelfoot Lake were created and nurtured by this river. Everything you see for miles around is in the Mississippi River floodplains; it's almost too big to contemplate. But, it's no longer a natural river; it has been redesigned; it is still part of a Manifest Destiny to conquer instead of complement. There are surely better alternatives for managing this river. I'll name one: restore it. We can start by restoring West Tennessee's tributary rivers. Self-maintaining rivers cost little or nothing and provide a yet-unidentified and unlimited list of major benefits to a nation. Man-controlled rivers are very costly, destructive, and unhealthy; and although they provide valuable benefits, these are limited and narrowly focused to address a few industries.

Congress must know that the Mississippi River has far greater values to consider, not least of which would be to restore the nation's native wetland deficit. It's not that the current industries benefited are unimportant, but alternatives have not been weighted and the tremendous benefits of former native rivers have been ignored. We have to ask the question: is the US Corps of Engineers the final word on how the nation's rivers and wetlands are to be used and managed? Or is it Congress who gives them the authority?

Reelfoot State Park

The park is the epicenter of outdoor recreation and conservation education at the lake. State Park headquarters is at the Ellington Center on the south shore of the lake. The Great River Road connects the park to twelve other Tennessee parks, wildlife, and recreation areas along the Mississippi River. Almost three hundred acres are set aside for campgrounds, interpretive centers, and trails, and facilities continue to be upgraded. The center has an excellent museum, boardwalks, and a nature center. Recently the park has added new module units for lodging at the Spillway. Bald eagle tours are popular during January and February. The park has two campgrounds: the Air Park Campground (fourteen campsites), at the north end of the lake, and the Blue Bank Campground (eighty-six campsites), at the south end of the lake.

Air Park Campground has about two miles of loop trails through mature stands of cypress along the lakeshore and mixed old field habitat. These trails continue to improve and offer a lot of diversity. The trails offer a nice outing any season, but require hiking boots or low rubber boots winter through early spring. Keystone Park has a one-way trail along the lake shore. The staff offers wildlife tours, interpretative programs, hikes, pontoon boat trips, a suggested self-guided auto tour through the swamps, and special events like a guided Deep Swamp Canoe Float, a Jr. Ranger Camp, a Pelican Festival, and other great outdoor programs.

Chapter 8

Solitude, Changing Times, and Lore

Hollywood and the Cranetown Rookery

The solitude and scenic beauty of Reelfoot Lake did not go unnoticed. Hollywood found it. Several movies have been filmed at Reelfoot. One of the earliest movies was the 1951 production of *Rain Tree County*. Something in the director's creative mind required the production to be filmed in the haunting excitement of the Cranetown Rookery. Unfortunately, the caution we take today for the security of nesting wild birds was not a serious issue in the 1950s. I'm told that they were not shy about making noise—even with the use of dynamite!

A heron/egret rookery is anything but a quiet place, although it is certainly a sound in nature you will not soon forget. One of the sets for the movie was practically in the middle of nesting great blue herons, egrets, anhinga, cormorants, night herons, and a few other water birds. Oddly enough, in 1919 and 1921, not a single American egret was found in the rookeries of Reelfoot Lake. But, by 1932, approximately 450 nests were reported in Cranetown. (1) But the entire rookery was abandoned in 1956. Widely reported, however, was that the rookery had one thousand nests in 1938; by 1961, not a single nest was to be found.

Rookeries naturally relocate from time to time for a fresh start. Yet, old timers report that the activities of the movie-making in 1951 were the direct cause of the abandonment at Cranetown, which stressed and scattered the birds. But these were also the years when the pesticide DDT was found to be mortally destructive to many species of wild birds. So we cannot be certain that the disturbance of the movie-making alone was the cause of the abandonment. By the 1970s, however, a new colony had been established at Little Ronaldson Slough, not too far from the original. I have observed this colony many times, and it appears to be doing well.

A more recent movie-making event was *U.S. Marshals*. Part of that

Figure 26. *U.S. Marshals* movie set;
night chase for Wesley Snipes.

movie in 1998 enlisted a dozen or so Lakers as extras for a week or so of
filming on Grassy Island. Much of the filming could be observed from the
boardwalk and observation tower at the end of the Grassy Island Road.
The scenes were drama-filled, but added nothing to show the true character of the lake or its people; rather, it was Hollywood-serving film: a film
with dismal heroes, a "dismal swamp," and dismal backwoods swamp
people. Some movies might fit a natural area; some don't.

Reelfoot Lake has long been known as a place in which water lilies, cypress groves, hidden basins, remoteness, and solitude were easy to find.
A splash from oar blades; the smell of pan-brewed coffee, skillet-seared
coot gravy at 4 a.m., the aroma of frying fish—they're part of Reelfoot, elements of nostalgic memories. There's others, like the sound of trammlers
running fish into their nets: hear the "thump, thump" of a push-pole on
a wooden boat to run the big catfish, buffalo, and carp into the nets. Hear
the heavy splashes of a trammeling block trashing water—sounds once
so commonly heard from skiffs captained by fishermen like Red Basham,
James Scheland and Lonnie Bratton, or Wendell Morris and Terry Pride.

And off in the distance was the common baying of someone's lost 'coon dog, still hot on the trail of something in the depths of a cypress swamp. These were a few of the common sounds one might hear at Reelfoot Lake. Somehow, none was the slightest bit intrusive. You might still hear them, but these nostalgic sounds are becoming scarce.

"Click-clack, click-clack" was the quiet sound of oars on a stump-jumper coming from the lake. You rarely hear that sound these days; not the baying of 'coon dogs, the bump of push-poles on wooden boats, or the sounds of a trammeling skiff. While we can still find quiet and seclusion on a good day, the old ways of quiet oar boats have surrendered to the whine of outboard engines and flat bottom aluminum boats. But that day, too, is yielding to another era: 110-hp bass boats, more suited to Lake Okeechobee or Kentucky Lake, now skid along the bayous, ditches, and canals. You might find the modern upgrade of powerful engines and speeding boats a bit annoying, but we have to live with those, too—unless we agree to change it.

A line in the sand has been stepped across—speed and obnoxious human noises are anathema to the nature of Reelfoot Lake—it is a natural area, not a carnival or racetrack. Thank goodness for stumps, or we would have ski boat competitions; Dollywood is a delightful place, too, but hardly a substitute for state park picnic grounds. There is a place for all of these things, but natural areas have their own limitations of use, or they are ruined. We have reached some of those limits. I have one fad left to address—airboats.

Figure 27. *U.S. Marshals* movie set; Laker movie stars.

Air boats are useful and powerful boats in their place but scourges of peace and quiet on Reelfoot Lake. I see the purpose of these boats as emergency or management watercraft. I had no part in its liberal use at Reelfoot. Actually, I made a formal protest against the recreational use of this watercraft to the TWRA Commission. It was an unfortunate luck-of-the-draw; a member of the commission happened to own a recreation boat company—and the answer was "Forget it." Airboats have no practical justification as recreational watercraft at Reelfoot Lake. No commercial justification, either. We are talking about airplane engines with airplane props. They skim the water and leave a blinding tornado behind—filled with debris, water vapor, bird nests, frogs, coke cans, coots, herons, and rails. It flies through cutgrass and fields of lotus and smartweed like a mad machine; it even roars across solid ground. The only thing that stops an airboat is a tree that refuses to bend. Any critter within a mile dives, goes underground, flies—or is chopped up. As soon as a happy airboat pilot (with ear muffs on) cranks the engine at 4:30 a.m., sleep is finished for the neighborhood—wildlife and people alike. Nerves are shattered. Complaints and threats are on the rise. Regretfully, a few fatal tragedies have discouraged local ownership (I lost a good friend). The machine has left its mark on more than a few skinned cypress trees, and some have ended upside down. Airboats are useful as emergency and management watercraft, but at Reelfoot, it is a "thumbs-down" recreation vehicle.

Mud Buddies, Go-devils, and jet boats are relative newcomers at the lake. Far safer, the decibels of these engines are about half those of air-boat engines. Pilots of these noisy boats—the only ones wearing ear muf-flers—are not the least bothered. These boaters probably have no clue they offend nature or anyone, probably because speed and noise fit the mores of these changing times, which have done little to upgrade the ethics of duck hunters. Few know anything about high-quality hunting, oar boats, and nature's solitude.

The ones we'd rather not live with are not so much the boats but the rowdy gangs in the boats. I call them the "Coco Beach-party" crowd. No sound but airboats more defiles the nature of Reelfoot Lake than this group. They seem to lobby against the lake for having no beaches. With-out a beach they simply drift a hundred yards or so offshore and begin baptizing each other with ferments, rebel yells, and "volume-up-max" radios. Their mission: convert the world to their inner circle. "Kill joy!" some say, but I will not relent; natural resources are treasures of the na-tion, where we should be humbled and glad to harmonize with its natural communities. And that point, well-known at Reelfoot Lake, and mostly honored, should be addressed as a hard policy, supported and enforced by lake managers and their agency administrators and commissions.

Who makes decisions for change? Lake managers and those who use the lake. But nothing happens until their concerns reach the ear of natural resource managers. While the public has had a lot to say about the use and management of Reelfoot Lake, large public meetings accomplish very little; rather, groups of less than half a dozen or so count the most. Personal letters and personal appearance are the energy behind public interest. Whether the petition is to commissioners, legislators, resource managers, or the CEOs of the agencies, with good facts and logic, the most effective communications are personal correspondence and face-to-face conversations.

The Times, They Are a-Changing

Like most of the nation's wetlands, the fragile ecosystem of Reelfoot Lake now faces many trials and threats. Future trends for things we love in these critical areas are not always secured in a changing world. Reelfoot Lake is not exempt: an invasion of exotic plants and animals, the careless use of pesticides and herbicides, excessive sedimentation from soil erosion, separation of the lake from its parent, the Mississippi River, and so forth. These discrepancies are more reasons users of the lake should take notice—be an advocate and remind government officials, wetland managers, and the media about your concerns. Natural resource managers are point men and women as watch dogs to remind us when and how we should prepare to avoid or adapt to these changes. But when trends of use are set by wrong policies and mismanagement, expect accountability. Even natural resource managers should be reminded that if quality management and equitable outdoor use is not part of their management philosophy, they are in the wrong field.

The mantra that "the most use of a natural resource is the best use" is from ignorance or disregard for the principles of good land stewardship. It does not stand up to the test of how much the resource can stand, how well the experience fulfills the needs of the human mind and spirit, or how much people will tolerate. Visitors and those who live here are point men and women, too. So, along with the joy of experiencing natural areas like Reelfoot, we must be compassionate about its care and future availability to those we leave behind.

What will be the primary uses of Reelfoot Lake in the coming years? Today, hunting, fishing, and wildlife observation remain the main activities at the lake. But, as I have suggested, the trend to hunt and fish is downward. So, if the next generation finds conservation education and watching wildlife to be the chosen sports, they must be given the same considerations to discover it. The natural beauty and opportunities at Reelfoot Lake will be with us for many more years, but the next generations will find it a changing ecosystem. Nevertheless, extraordinary outdoor adventures through non-consumptive uses will have much the same plethora of wonderful outdoor features to welcome its visitors.

As the lake ages, there will be new adventures. It will become more bottomland forest, more marshes and cypress swamps, and less open lake; a disadvantage to some creatures and advantage to others; less useful to some users and more to others. The next generations, with wise management, will see many more ways to enjoy the outdoors in non-consumptive fashion, more conservation education facilities like nature trails, boardwalks, and guided boat tours. Conservation and preservation should be on the minds of every nature-minded person. They will most likely want to salvage what little is left of native wetlands; they will accept the principles of nature to develop new wetlands—sustainable, and with features very near natural wetlands. If this attitude becomes the norm along the flyway, the downward trend of wildlife populations will reverse, and wildlife along the flyway corridor will begin to flourish.

Will the self-sustaining power of nature at Reelfoot be compromised? Not necessarily. Natural wetlands, undisturbed, are rare. It leaves little to compare with the tens of thousands of acres of altered or destroyed native wetlands like Reelfoot Lake. Wetland managers who understand them have been rare, but that day is passing; the science of wetland ecology has been revealed by good research and practical experience, and in that, hope. Natural resource managers have begun to understand past failures and now apply this knowledge. It will go a long way in the recovery of many of our losses, especially our former native wetlands, which support such a wide array of sub-ecosystem habitats, the sum of which are far more valuable to us than anyone but the most astute conservationists is able to imagine. The managers' toughest job is now is to convince their administrators, peers, waterfowl hunters, politicians, and the public the merits of their new techniques.

In vogue fashions can elevate or crash a clothing market. Fur markets at Reelfoot diminished when clothing styles around the world limited the use of wild furs. The first notable change was during the early 1900s, when beaver hides (used like currency during the nation's westerly expansion) were no longer in demand by European markets; silk hats became the fashion and fad of the day. Markets for furbearer pelts today have not recovered this ancient trade; and depending on this commerce has become a sparse means for a woodsman to even supplement a living. The commercial sale of nearly all wildlife now is very limited and has essentially ended as a means to make a living, although it played a large role in the livelihood of the Gumbooters. I am humbled that I was fortunate enough to have lived during this golden age at Reelfoot, but disappointed that I also saw the transition that ended the age of Gumbooter's best years.

Creature Lore, Facts and Fiction

Then, there was evening time. We sat quietly spellbound many evenings in the dim light of kerosene lamps and the din of smoke-filled rooms and screened porches, to listen while storytellers spun their lofty tales. Sometimes it was about creatures and things I never saw; the last panther, the last deer, beaver, wild turkey, and otter; they had been gone for a generation. Oddly enough, there was rarely mention of the lawless and hair-raising, real-life stories as close as their doorsteps; mayhem that made headlines during October 1908.

Rather, there was "Catfish Head."

Catfish Head was a larger-than-life creature, the mention of which silenced the teller's young listeners. The story said that the creature was an orphaned Gumbooter, raised in a hollow log by a mother flathead catfish. Catfish Head was not at all fond of trotline or snag-line fishermen, but he tolerated others. One thing for sure, you did not want to be on the raw side of Mr. Head; he could drown and swallow a person whole from the slightest provocation.

It took days, weeks, maybe longer for most of us to shake the images in these tall tales. Gone were endless nights of sound sleep. Thump, thump, thump . . . swoosh! . . . and the lake rippled like an asteroid landed. Logic said it could only be a huge beaver. But any dark night out on the lake could stir thoughts of Catfish Head. Many told of the nights he surfaced from black water at the side of their boat. They had disturbed the grave of his many victims. Witnesses had him passing through dark shadows along the Bayou du Chein, along familiar damp and dank footpaths that wend and wind their way beneath the tall groves of cypress trees. After all, it was said that a huge, hollow sycamore log sunken not far up the bayou was the creature Head's favorite home. We hated that rumor; that old tree was called "Aunt Ily's"—our secret swimming hole.

Our small troop of Huck Finn adventurers had a deceptive streak. Aunt Ily's sounds a lot like Aunt Ivy's, if you say it pretty fast. Aunt Ivy Henson, of course, was one of my favorite aunts who lived on the west side of the bayou. Aunt Ily's was an old hollow sycamore tree that stretched low and almost across the Bayou du Chein, about a mile upstream from our village. The very top of the tree dipped conveniently down only a couple of yards from the opposite bank of the bayou and back up—a perfect saddle for three or four boys to sit in after a frolic in the cool, shady waters of the bayou. Worried mothers, of course, did not approve of any such "foolish" excursions without adult supervision, which we definitely opposed. When they noticed us wandering away from the homebound territory, they would inquire where we were going. The answer (not to lie) was: *Aunt Ily's!* Was it deception? Yes, but our conscience was fairly clean; it always satisfied the inquisition. At the very worst, it was only a white lie. We had no intentions of going to Aunt Ivy's, but to our secret swimming hole—and maybe one of Mr. Catfish Head's abodes. I think we successfully kept the secret from our moms until we were adults.

There were many stories, like huge boulder-like objects, inexplicably dropped from the trees, causing monster splashes; massive dark shadows moved silently through the swampy bramble with penlights many nights in June; panthers screamed; some stalked our boats as we went along the bayou; some crossed our path on logs in plain sight; bright fireballs rolled across the road on moonless nights before witnesses. The tales were endless. No one will know the value of the times we sat spellbound on the plank floor of Mr. Grover McQueen's Grocery.

Chapter 9

Rising Conflict

Lake Ownership

The life of Lakers, as we have seen in fairly tolerable ways, has not always been so rosy. Living here during the early years was something of a hard life by modern standards. But they were pioneers with grit and they could contend with hard times. Commercial fishing, trapping, and a little gardening soon became enough for a comfortable life. Up until about the 1870s, less than a dozen commercial fishermen were available to supply J. C. Burdick's fish market at Walnut Log. Reporting as much as one-hundred thousand pounds of fish a month at peak delivery, residents did well in the commercial-fishing business. Quickly iced down, the daily catch could be shipped by rail to distant cities from Burdick's fish dock in Union City. Life had its inconveniences but they usually made it work.

Commercial activities collectively attracted more businesses and competition after the 1870s in the fledgling hamlet of Walnut Log, and the rising towns of Samburg and nearby Tiptonville. The outside world had discovered Reelfoot, and the region was soon a buzz of activity. The commercial sale of ducks became a regular source of income. Barreled and salted down for big city markets like Chicago, duck too was an important winter source of income. European fur markets sustained the art of trapping as another source. Sport fishing at Reelfoot was a sport for only a select few locals during the early days; everyone else fish mainly for profits trying to survive and take care of families. As competition for the lake's natural resources increased somewhere around the 1900s, personal and reliable hunting and fishing territorial grounds shrank because of competition. Discontent stirred among them. The solution took time but was tentatively settled by an unwritten code of respect for trapping and commercial-fishing grounds that had a tradition of individual territorial boundaries. It was actually "a code of respect" for each other, and it worked quite well for a good long while.

About the end of the first decade of the 1900s, the Laker's world was infiltrated by men of a different mindset who declared ownership rights and no longer cared to share lake resources—unless the user paid. Lake People were of a different mind; they did not care to share the lake with outsiders, unless the outsider paid. Those attitudes were bound to clash legally or physically, and they did. While there were honorable intentions, a caste of these men was inclined to be on the side of renegades, their intentions without neighborly concerns or consequences.

J. C. Harris was an ambitious landowner whose son Judge Harris later formed the West Tennessee Land Company. Harris was the greatest single influence on the future of the lake at the time. Through an extensive North Carolina Land Grant and ingenuity, Mr. Harris owned most of the land in and around the lake. Eventually he intended to own all of it, and he left no doubt as to his objectives: the lake would be drained to expose suitable farmland, and cypress swamps would be drained enough to extract the virgin forest for lumber and have it hauled to local markets. J. C. Burdick, although he would become a partner to the West Tennessee Land Company, enjoined Harris in 1899 from draining the lake. Harris died in 1903, never accomplishing his objectives.[1]

J. C. Harris left his estate to his son Judge Harris (a given name). Judge had little sympathy for the Gumbooter's philosophy of ownership and required pay if the Lakers used the lake. Judge was also a progressive thinker, and he thought corn crops would do more good for society than any other use of the lake. Eventually, he would own land under the entire lake and most of its shoreline, save only a couple of tracts. John Shaw of Samburg bought one of the disputed tracts. Shaw, Burdick, and Pleasant (all from Samburg), and P. C. Ward of Walnut Log joined Judge Harris and formed the West Tennessee Land Company, the caveat needed for ownership of Reelfoot Lake. Eventually, Shaw and Pleasant withdrew from the partnership; locally, they were considered traitors.[2]

Reelfoot fishermen saw their occupation put in jeopardy—the golden age of living at Reelfoot Lake seemed to have passed them by. Whether Judge Harris intended to drain the entire lake and farm it, as his father did, was not certain. But Harris, still the primary owner of the company, was a business man, and making money was his business. Wetlands to Judge were no more than mosquito-infested swamps, and so were the minds of most back in those days. Even today, the stigma of "draining the swamps" has a positive effect in most circles. But ask an avid duck hunter today the value of a natural marsh; most can't afford the price tag.

Judge Harris's plan to profit from the commercial use of the lake, as could be expected. His ambition caused a local uproar. Lakers found no legal way to get Mr. Harris to change his mind. To make matters worse, lawyers hired to defend their interest became officers of Judge Harris's West Tennessee Land Company.

Desperation set in. The plaintiffs appealed for relief from Judge Coopers' court in Trenton. Again, the Gumbooters were rejected. Now, they were even more desperate. For some there was simply nothing else to do but hope, so they placed their future on the courts. Others grew impatient; the courts had so far done nothing to give them relief. Nevertheless, as a group they had no intentions to back down. The lake was the only life they knew. It was a drastic decision that reset the future of Reelfoot Lake.

A one-sided deal was made. J. C. Burdick Reelfoot Fish Company and other commercial-fishing dock owners, knowing they were beaten, contracted with the West Tennessee Land Company, which gave them exclusive rights to the commercial take and selling of fish and game from the lake. It didn't last. Partly because the West Tennessee Land Company required a tax from its commercial-fish dock owners, fishermen were paid less for their catch in order to compensate the tax. Not a deal for most of the fishermen, and they refused to accept the terms. From here, the history is messy.

Night Rider Mischief at Walnut Log

Unable to afford credible lawyers, a band of justice-seeking desperados was formed to defend local interests. Some were fishermen, and some were some farmers from the hills (the primary leaders), and some were both farmers and commercial fishermen. These were the Night Riders, some with good intentions, some questionable. The Riders developed their own creed and brand of justice; they meted out punishment, sometimes it seemed justified, sometimes immoral and self-serving.

Local support for the Night Riders (who operated under the cloak of darkness) was initially favorable to the cause. The popular view was that the Riders sought some way to convince the West Tennessee Land Company Board to rescind contrary decisions affecting them, but their methods eventually went awry. Being judge and prosecutor for their own brand of local justice, they became involved in killings, whippings, and burning of some homes and businesses—for what the Riders considered wrongdoings. Some victims did not even know why they were being punished.

John Shaw was whipped with thorns and warned not to tell what he knew. H. B. "Con" Young ran a sportsman's lodge of sorts on Starve Island, a small tract of some fifty acres on the lake. He was given ten licks with a rope and told to get off of the island. "Con" Young complied.

The Riders overstepped their legal boundaries exceedingly in October of 1908. Their undoing was when they thought two of the company attorneys, Captain Quentin Rankin and Colonel R. Z. Taylor, could be coerced to provide relief from the West Tennessee Land Company policy that harmed them—or be hanged. They were wrong.

The attorneys had lodged at Ward's Hotel on company business. Sometime after dark, with some thirty-five members of the clan, the Riders entered the lodge. Unsatisfied by the attorneys' lack of cooperation, they marched them back up the road to a location near Mr. Ed and Mrs. Myrtle Powell's house (my family once lived in the Powell house; it was also the future front yard of my parents' house when the old Powell house was destroyed).

From a tree on the bank of the Bayou du Chein, they put a rope around the neck of Captain Rankin and repeated their demands. Filled with anger, the Night Riders hauled the captain up by the neck more than once.

Figure 28. Walnut Log or Ward's Lodge, 1908. Built on floating logs. General Taylor and Captain Rankin were taken from here the night of the abduction.

Figure 29. A newspaper photograph
circulated widely in the Reelfoot Lake area
concerning murder by the Night Riders.

Unable to convince the captain, they riddled him with bullets. Sixty-three year-old General Taylor dove into the bayou and hid under a log on the far side. Spraying his location with bullets and shot, the Riders thought they had killed both of these men; Rankin's body lay where it had fallen.

But General Taylor miraculously survived the barrage of gunfire. Afraid to stay in the Walnut Log area, or Obion County, where most of the Night Riders originated, he headed westerly in the dark, where a couple miles of swamps and wilderness lay between him and human habitation, jungles I knew to be of the toughest kind. I would hate to follow his trail with inadequate clothing and sustenance supplies. He spent a restless night somewhere out of reach of the Riders on a spot of dry ground to assess his options: If he crossed the head of Upper Blue Basin, he would go through some of the meanest swamps you can imagine; thickets with cat-briers as sharp as razors, and giant cutgrass infested with cottonmouths; it would be a slow and hazardous route. If he went north, he could follow a shallow ridge on the west side of the Bayou du Chein all the way to Kentucky and an east-west road. In October, the forested ridge is fairly free from standing water, but a favorite ground for denning cottonmouths. Even so, it would have been the general's best route to avoid the worst of the swamps. Either way, in the pitch black of dark, he had no safe or easy route to reach a living soul for help.

Sometime the next day, I believe, General Taylor arrived, torn, hungry, thirsty, and exhausted, at Luther Rankin's farm in Lake County. No kin to Captain Rankin, Taylor knew Luther as a friend. The ordeal was over for General Taylor, and so it would soon to be for the reign of the Night Riders. The crisis resulted in several court cases, in which many of the clan were found guilty of murder. Ultimately, however, none was prosecuted. Despite the cruelty of the Night Riders, a solution to the conflict soon followed—the state bought a strip of land around most of the lake and land beneath the lake as public domain; the state park was named Reelfoot Lake.

While details of this story have been circulated and told over and over around Reelfoot Lake ever since I can remember, the best written accounts are by two authors I have known: Paul J. Vanderwood and David G. Hayes. Paul J. Vanderwood's book *The Night Riders of Reelfoot Lake* is specific to this historic event. He goes into minute detail about the story as a university thesis. David G. Hayes's book *The Historic Reelfoot Lake Region* adds even more to the story through a historic perspective of the region. Hayes was born and raised here, and as an attorney, he has unique, personal, and professional insight into this history.

The period of painful regrets and healing was slow. Gray clouds of guilt and investigations hung over the residents of Reelfoot Lake for several years. Who was to blame? Seems nearly everyone living here had a role to play. But the business ambitions of the Harris family leadership and the recklessness of the Night Riders' reign of 1908–1909 bore responsibility.

The debate over the laws of navigability was of great concern in determining these rights. The state supreme court sided with the West Tennessee Land Company that most of the land beneath the lake was private property, and so was the lake if all of the land beneath the lake could be owned under the same title. Four years later, in 1913, the state supreme court reaffirmed that, indeed, the private deeds beneath Reelfoot Lake were valid. Yet, the lake was also of great value to the public. Considerable interest by the public supported this view, and the state gained great public support to purchase the lake. Finally, the state did purchase the land owned by Judge Harris and bought the rights of the Reelfoot Commercial Fish Company. The West Tennessee Land Company received $25,000 for its share and J. C. Burdick, $2,000 for his commercial fish business. So, for the bargain price of $27,000, fourteen thousand acres of land was turned over to the state in October 14, 1914, under the name of Reelfoot Lake.[3] Finally, the golden age for life at Reelfoot Lake had arrived.

Chapter 10

The Mississippi Flyway

The Cotton Patch Kid and Fall Migration

When cotton patches turned snow-white, most other activities around Reelfoot Lake were on hold. Schools closed the doors for weeks, and kids were found pulling cotton sacks, sometime as late as December. By this time, a rush was on to have cotton picked before Christmas. Education was not lost during this brief hiatus: cotton picking was a powerful extra-curricular course—if picking cotton wasn't your thing, you'd better finish school. It was hard work, but I didn't much mind it; fall was a magical season, a wonderful time to be outside. Any excuse to be out of the house or classroom was a good thing. Besides, full cotton sacks bought school clothes—and the fall waterfowl migration would begin.

But what kid was disciplined enough to keep their head down in cotton stalks with so much fall commotion going on in the clouds? The fall migration of waterfowl would be returning from the north; that was my diversion from the hum-drums of cotton picking. It was impossible to ignore them. Flight after flight soared beneath the clouds, day and night, all headed south, loquaciously conversing with their beautiful chatter, cackles, and song. Like many things in the world of Gumbooters, all that began to change with modern cotton-picking machines, for one; the downward trend of migratory bird populations, another.

But during my youth, I never anticipated such things. It was just cotton-picking time. And to me, it was synonymous with fall waterfowl migrations. There was a patchwork of fields around the lake beginning in September that looked like mile-wide prairies of this fluffy stuff, white as the topside of summer clouds. Trailer loads sitting at the end of the fields appeared as inviting as a mother's comforter. My entire family was involved during some of these years. As a teenager, I already knew fields like these inside out; I helped plow them with mules, planted the seed,

Figure 30. Cotton bolls in late September display their soft, fluffy beauty.

and cultivated the young plants with the same mules. Along with rest of the family, I sometimes weeded and spaced these tiny okra-like cotton sprouts for endless miles with a garden hoe. By early fall, I began to think I was little more than a "cotton-pickin'" kid, which wasn't the true calling of a "Son of the Swamps."

Welcome ducks! From late August through the coldest winters of January, we watched their migration, mostly new broods fresh out of flight school, family flocks that began their journey at their northern brooding grounds before heading south. Without looking up, I could often call the flocks by name merely by their vocals. Mallards and gadwalls chatter or quack; widgeons and pintails whistle like piccolo flutes; blackjacks and scaups, a raspy attempt to quack; and so forth. But a dozen or so of their wingmen-cousins, like teal, canvasbacks, redheads, and spoonbills, hardly say anything. They are such "speed-dos" all you hear upon their arrival from the heights of clouds is a sonic screech of wind through wing feathers. But the rest—the mallards and other puddle ducks—seem always to be conversing happily. Maybe it concerns navigation, stopover schedules, or perhaps the simple joy of flying. Flock after flock crossed the skies, tens of thousands in a single day. At night, they were more subdued. The bright but lonesome calls of geese and the quiet chattering of mallards, I still hear in daydreams when there are empty skies, and in the evening I still imagine their silhouettes crossing the face of the moon.

Figure 31. Gossiping Snows, Blues, and Canada Geese on Black Bayou Refuge.

Skeins of gossiping geese, mostly snows, and blues, and white-fronted flocks, have their own flight plan. Some fly in "V" formations, even higher than the ducks, one behind the other in continuous waves. The sounds of waterfowl could be heard any time of the day back then. Most of the tundra-born geese rarely stopped here when I was a kid, but that began to change about the year 2000. Like ducks, their orchestrated cacophony seems to crescendo to celebrate at the mere sight of Reelfoot Lake. Happy ducks and geese they were; I could tell. At some point, tundra geese decided that Reelfoot was an acceptable winter habitat, but not so much when I was a kid. At present, it would not be too ambitious to estimate nearly a million snow and blue geese (they are the same species) in and around the lake on certain winter's day. Most of these sights and scenes, my young mind in the cotton patch thought, would be here forever. And why not? Their presence should not be left to daydreamers.

From the cotton patch, it was only a short walk through the woods to the lake. I was tempted to hide my cotton sack and head through the woods to the lake. But I learned to be accountable and stayed. Mesmerized, I watched as a single flock could not resist an invitation from the lush marshes below. They broke, one after the other, from large formations. How graceful as they peeled off out of the northern sky in smaller flocks, wings set in tight formation. Suddenly, and on cue, an acrobatic maneuver; a formation makes a quick dissent toward the open lake;

Figure 32. Mississippi flyway map.

wings screeched from speed as they strafed the marsh. Unrestrained enthusiasm finally subdued, the flock splashes down at a lodging site of their choosing. Sometimes a flock would stay several days, but prudence was in order; during hunting season, the choice had best be a refuge, unless they came after the waterfowl season closed. But the times, the times have changed. Today, I sometimes wonder if I'm still on the same planet; their numbers are down, their habits have changed, and the skies are nearly empty of these migrating flocks.

The lake itself has changed. For reasons I've yet to understand, the great fields of giant cutgrass potholes were once favorite places for ducks to roost. Just at a certain candle of light, the sky was filled with puddle ducks seeking their favorite pothole to spend the night. But just before the ducks arrived, were the blackbird flocks—thousands upon thousands of them. Like wide blankets of mottled black velvet, sometimes hundreds of feet wide, the birds swarmed gracefully in undulating unison above the grass fields. By and by, they would begin to settle in a favorite place and delight in much chattering conversation until well after dark. Today, I sometimes wonder if I'm still on the same planet. The numbers of black birds and waterfowl are way down; the skies are nearly empty of these migrating flocks.

Of course, I was biased from my earliest experiences, but I believed without advice from ornithologists that the Mississippi Flyway was the greatest migrant bird flyway on earth. I could not have been more right, because it's true, and Reelfoot Lake lies midway between its north-south extremes. Shaped like a fetus soon to be born, its brilliance the color of life, comforted with the care of its river mother's bosom. Isolated only a few miles from its parent, the Mississippi River, Reelfoot Lake lies as an island oasis for children of the wild.

Reelfoot, one of four major migratory bird flyways in the Lower 48, is in the Central Flyway. Free from hills or mountains, the flyway follows the corridor of the Mississippi River some two thousand river miles in its shortest path from its headwaters in Lake Itasca, Minnesota, to the Gulf of Mexico. Mark Twain concluded that if you add the Missouri, the Mississippi would be four thousand miles or more—the longest in the world. Of course, the Missouri must be added since all of its tributaries plus the main channel make up the Mississippi River. Mark Twain mentioned it has fifty-four tributaries navigable to steamboats and hundreds more smaller steams. Imagine a lake with a parent river that drains a watershed greater that all of Europe. That's the Mississippi. Only the Amazon is larger in extent.[1] That's Reelfoot Lake.

There were reasons other than my bias that the Mississippi Flyway is the greatest river on earth. It has the fine distinction of having the greatest quantity and number of species of birds in the world. Approximately 60 percent of North American birds migrate through the Lower Mississippi. Some birds migrate from as far as the Arctic Circle to join this flyway on the way to the Patagonia in South America. Forty percent of waterfowl and 325 bird species are known to migrate from their summer nesting grounds in the north along the Mississippi Flyway to their southern wintering grounds; more than two-thirds of them winter in the wetlands of Louisiana. But there has been a change in the species of birds in this flyway, and all of them affect birding at Reelfoot Lake.

Table 1. Bird species present in the Reelfoot Lake area in 1810, but now absent.*

Trumpeter Swan	Common Raven
Swallow-tail Kite	Backman's Warbler
Whooping Crane	Passenger Pigeon
Ivory Billed Woodpecker	Carolina Parakeet

Table 2. Bird species absent at Reelfoot Lake in 1810,
but now present.*

Cattle Egret	Starling
Rock Dove	House Sparrow
Mockingbird	House Finch
American Robin	Blue Grosbeak
Dickcissel	

*Johnson et. al., *Reelfoot Lake 50-Year Plan, 86, 1988,*
Tennessee Wildlife Resources Agency.

Here's the kicker: Reelfoot Lake is a premier wetland habitat for birds, a smorgasbord for migrants going and coming, an "oasis" among the wetlands found intermittently along the river corridor. Like a "string of pearls" for migrant birds, each place is sacred to their future. The migratory bird population in general trends downward—all the more reason to emphasize the need for quality habitat throughout their range. The value of these remnant wetlands is so critical to the survival of fish and wildlife and migrant birds, I dare try to describe it. Along the way south, the river collects hundreds of daughter flyways, like some I just mentioned. These river corridors also deserve praise for being ideal navigational pathways, because they collect migrants from two-thirds of the Lower 48 states and major parts of Canada.

Cheerful are the days we hear the waterfowl and shorebirds call from the heights of clouds. We rarely hear the tiny warblers flying so high, but we can see them flit from limb to limb through tree canopies, and hear the buzzes and cheeps of other birds (some of which I no longer can hear) fill the canopies. From cotton and corn fields to city suburbs, the sound of their melodious notes fills the canopies every spring and fall. Look up to admire the V-shaped flights of geese, skeins of chattering ducks, or hear the sweet sounds of neotropicals. What comfort it is from our scurrying about below to know that part of nature is often a mere tilt of the head upward to the treetops—or somewhere a little beyond. Reelfoot is one place to experience these things.

Yet the modern imagination of artificial worlds sometimes threatens to take the edge off our naturalist way of thinking: there is a rumor that these are not really birds we see but government drones to spy on us. If so, they should surely wonder why we think so highly of a traditional holiday dinner; roast "drones" and cornbread dressing at Christmas! Pity, I say, that these deficient cerebrals have need of a learning adjustment by a short winter visit to the wildlife viewing towers at Black Bayou and Long Point refuges during the peak of migration, or read the comforting works of naturalists like David Thoreau, John Muir, and Aldo Leopold. That's why we will always need Reelfoot Lake—to learn about nature and the need of it as well as outdoor recreation.

Why do migrant birds have such great affinity for the Mississippi River corridor? The best answer is not the absences of hills and mountains, but, diminishing as they are, the greatest abundances of *native wetlands*. That alone is reason enough to appreciate these mantras:

> Rivers are the mothers of most native inland wetlands.
> Freedom for the river to flow is a necessity for the creation of new wetlands.
> Without the freedom to flow, new wetlands are not formed and the future needs of migrant birds are cannot to be met.
> The Mississippi River was essential for the creation of Reelfoot Lake.

They all are true. Fortunately, we are beginning to learn the full meaning of these mantras. If for no other reason, Reelfoot Lake should be on a state priority list of critical needs to learn how to manage our diminishing wetlands, not to dismiss the dire need of native wetlands for wild things, people, outdoor education, and the integrity of Tennessee's beautiful landscapes. Changes are on the way, and vocations at Reelfoot Lake will need to adjust to meet this new era. That will very likely require a new initiative with compatible regulations, topnotch trails, guided tours, and boardwalks galore.

Waterfowl is almost synonymous with our four flyways, although many other migrants use the same flyways. Major waterfowl flyways usually overlap with other bird flyways. Interest in waterfowl populations is extremely important nationwide to some two million waterfowl hunters. Since they also buy licenses for their sport, state fish and wildlife agencies have a special interest in waterfowl management. That's the major reason that Tennessee Wildlife Resources Agency and US Fish and Wildlife Service emphasize waterfowl refuges—major support comes from waterfowl hunters.

Reelfoot is an oasis for migrants, a major stopover for waterfowl, going south or north. Largely winter visitors at Reelfoot Lake, wood ducks and giant Canada geese nest and raise broods here, and are found here all year. Migrants arrive after a long summer on nesting grounds, incubating eggs and tending to busy broods searching for bugs and avoiding predators. Growing family flocks stay close to their nesting grounds until time to leave during the fall. With good health and good luck, the family anxiously awaits the day when their internal clocks signal departure time, time to move to southern climes where new fall crops lay waiting to provide a feast. They leave with great excitement and expectations that special wetlands like Reelfoot will be here to provide backup security through temporary lodging and food to help provide traditional needs. Here, they rest a bit before going farther south, some even stay the winter, and some go on to South America. This is where the broods and their parents escape the harshness of winter and find suitable habitat; where the young discover adulthood, and return to their brooding grounds to replenish the flyway another year.

The entire twenty-four thousand acres of Reelfoot Lake is prime waterfowl habitat during some months of the year. Through flooded forests, well-maintained marshes, and managed cropland, we are optimistic that the success of waterfowl populations continues. The Long Point Unit of the Reelfoot NWR has long been the index unit for waterfowl populations at Reelfoot Lake. Duck populations from 1997 through 1980 averaged a peak of 261,738 ducks; between 1983 through 1986, the average was 116,372. What this says about the abundance and quality of habitat is influenced greatly by droughts and wet seasons in their production grounds, but there is little doubt that their waterfowl wintering grounds have declined.

Eventually, I learned that ducks came mostly from the nesting grounds of prairie potholes in the near northern country and from rivers like the McKenzie, and farther north to Alaska and the Arctic Circle. Snow and blue geese came from nesting grounds in the Hudson Bay vicinity, and white-fronted geese mostly east of there as far as Alaska. Oddly enough, few Canada geese wintered here when I was a kid, although a captive flock was present at Long Point Refuge during the 1950s. The purpose was to encourage these geese to winter farther south than Horseshoe Lake in Illinois. It worked really well . . . for a while.

Harsh winters and refuge preparations at the Long Point NWR Unit held great populations of geese in the Reelfoot area from the late 1950s until the early 1980s. During the 1985–86 year, the refuge held some seventy-thousand Canada geese. Fifteen years later, one could hardly find a common Canada goose (*Branta canadensis canadensis*) anywhere in the region. Today, migrant Canada geese have virtually been replaced by their cousin giant Canada geese.

"Newcomers," known as giant Canada geese (*B. c. maximus*), are larger than common Canada geese. TWRA captured and relocated many of these geese from Old Hickory Lake near Nashville to farm ponds and small lakes. These geese prefer to graze low-growing grass around water, like golf courses and manicured yards around large lakes. Sadly, it gets them into a public relations problem. I was surprised that anyone would complain, that these magnificent birds would ever be called a "nuisance," cursed because they picked a little lawn turf and left droppings. Worse yet, they had the audacity to do the same thing on country club golf courses. Like whitetail deer and wild turkey, there were no giant Canada geese to be found in the wild in this region for half a century. Fortunate for those who need to control irksome wildlife, wildlife laws are in place to allow legal control.

But "newcomers" they are not. Famed artist and ornithologist John J. Audubon and others have noted the sightings of these geese in their journals on trips along the Mississippi near here. Since then, the world of waterfowl has changed in the flyway. Presently, duck flocks are down but it could be from shrinking prairie potholes, which are very important nesting grounds for waterfowl. Today, the current trend is dwindling waterfowl-nesting wetlands, and they will find the trend more so for a favorable winter habitat. Oddly enough, giant Canada geese, unknown in West Tennessee by the 1900s, are nesting on the lake today.

Even today, nothing elicits quite the feel of the wild as the sounds and sights of fall flights of geese and ducks. Our good fortune is to have the continued presence of giant Canada geese to remind us their cousins should be here come fall. I have vivid memories of the days when out the backdoor of my house was where duck-hunting memories were made. It took about fifteen minutes with hip-high gum boots to get to the giant cutgrass potholes my cousins and I hunted. According to my father and his father, it had been that way since they could remember. I knew very little about the ecology of waterfowl; I simply expected they'd show up every year. But during the winter of 2020–2021, the populations of wintering waterfowl were quite bleak. Their future all depends on us: can we take what measures we know will help quickly enough to assure they will be here next year?

Like giant Canada geese, several other residents were here until the mid-to-late 1800s, but most had already vanished and the rest by the turn of the next century. By the early 1900s, whitetail deer, wild turkey, otter, black bear, elk, and a few others had been extirpated or were very rare.

Whitetail deer were gone from the Reelfoot area by the 1880s. My grandfather said he shot the last deer he knew about. They were absent until 1935, when twelve whitetail deer were restocked in the State Woods. The first quota hunt was 1978, when three bucks were harvested. Wild turkeys were stocked in 1941 by the Game and Fish Game Farm; but nature had news for us: only wild-captured birds could survive and replenish depleted populations by restocking. Only wild parents can teach their offspring how to survive. So, thirteen wild turkeys were captured by canon nets from Shelby Forest in 1961, and seventeen more in 1966, which is why wild turkeys are common at Reelfoot Lake today. Successful wildlife restoration is an example of the strides modern wildlife management has made since it began in the 1940s with biologist/forester Dr. Aldo Leopold, who wrote *Sand County Almanac*. TWRA currently has long hunting seasons for all three of these game animals.

Some birds get lost in the name calling. Many kinds of waterfowl, and wading and shorebirds, have long been legally hunted. Snipes, woodcock, and rails are still legal, but most of the shorebirds and wading birds are now on the list of protected species. My father and grandfather often spoke of harvesting cranes, of all things. "Gordhead cranes at Reelfoot," they said, "were quite sporting, plentiful, and a favorite bird for table fare." What were gordhead cranes? It would be years before I knew. But cranes (local for any kind of heron or egret) were sacred in my mind. If it didn't act like a duck, and quack like a duck, I had no interest in shooting it. Never knowing what kind of bird they talked about, since none was left by the time I was a kid, I remained oblivious.

Gordhead cranes, however, were not the only "cranes" to be hunted in America (nor at Reelfoot Lake). The greatest commercialization of wading and shorebirds was during the early 1900s to satisfy the feather-plume styles of women's hats. I have a few friends and acquaintances that have relished the breasts of great blue herons, and some owls, starlings, blue jays, and others with any size. The Migratory Bird Treaty Act of 1919, however, is very fussy about which ones are legal today, and the wrong choice could be very expensive. Bird feathers are indeed, beautiful, but they look much better on their original owners.

I found gordheads to be sandhill cranes, close cousins of the endangered whooping cranes. Standing a little taller than a three year-old kid,

Figure 1. Reelfoot Lake: An oasis on the Mississippi.

Figure 9. "I found the Lake aglow with Yellow Lights of the Great Lilies."

Figure 22. Black Bayou at sunset.

Figure 24. An infrared photograph of Reelfoot Lake (dark red is trees or heavy, green vegetation; blue is open agriculture land; Kentucky Bend is in the upper left corner).

Figure 25. Foggy morn at the
Ellington Center Boardwalk.

Figure 39. Green tree frog hidden in lotus bloom.

Figure 40. Baby painted turtle.

Figure 42. Muskrat sunning, Horse Island Canal.

Figure 43. Gold finch on wild thistle, button bush, and prothonotary warbler with green worm.

Figure 49. This noisy juvenile bald eagle entertained us
the summer of 2018 begging its parents for food.

Figure 54. Soft shell turtle pair on log. (It took stealth
and a 400-mm lens to get this shot.)

Figure 55. Sunrise and sunset are wonderful times on the lake.

Figure 60. Great blue heron tries to avoid spines of catfish.

Figure 66. November egrets go to roost.

Figure 67b. White pelicans, Brewer's Bar.

Figure 74. Mr. "B" runs trotlines out from Gooch's Landing, Lower Blue Basin.

Figure 75. Cottonmouth: The only poisonous snake known at Reelfoot Lake.

Figure 77. A tiny green cricket frog in duckweed.

Figure 83. Ancient bald cypress trees displaying the bronze beauty of their fall leaves and a time-worn skeleton.

Figure 33. Gordhead cranes return to Reelfoot Lake, 1990s.

these birds by the 1930s were hunted to extinction in West Tennessee. Dr. David Pitts, University of Tennessee, Martin, stated in a 1982 Reelfoot Lake study that only one recent sighting had been reported in West Tennessee. As life would have it, I would one day play a role in the restoration of these birds.

About 1993, TWRA had just finished the final stages of the state's newly developed Hop-in Refuge on the Obion River, when Ralph Gray, the WMA manager, and I stood at sundown to admire our work. At the last blood-red light of sunset, we were alerted to the melodious calls of Sandhills—I had never heard or seen them in West Tennessee. But I knew instantly what they were since I remembered their melodic calls from somewhere. As if by pure magic, half a dozen of these handsome birds crossed the setting sun, noisy, with wings set to land on our new refuge.

I took a picture of them as they approached, but the light was low and so was my film speed; it was a lousy shot but I still look at it from time to time. It was one of those rare and memorable rewards one can get while working with nature. The flock is today more than two thousand birds, and I have heard sandhills pass over Reelfoot in small flocks even though I rarely see them. The first open-hunting seasons for the Hop-in Refuge flock, I'm told, began this year. We hope to find sandhills a regular visitor to the Reelfoot area in the near future.

Make no mistake thinking these wetlands I call Pearls of the Mississippi River are only for waterfowl and waterfowl hunters. Far from it. By their own natural clocks, a long list of other birds that follow the Mississippi Flyway make the same ancestral journeys every year: tiny warblers, bald eagles, osprey, white pelicans, shorebirds, bitterns, herons and egrets, swallows, finches and buntings, hummingbirds, and a dozen others migrate from the northern forests' nooks and crannies. They can be enjoyed right here at the lake. All that's needed is time, timing, and sometimes boots or a boat.

Birders are special visitors at the lake: birders come quietly, take nothing, and leave the same way. All they leave are footprints, and I can't say I've found many of those. We have no good method to determine how much use the lake gets from birders, since they buy no licenses, or any equipment to tag that might reveal their use or contribution.
Short of special user surveys and membership in organizations like the National Audubon Society, their numbers are estimates at best.

America drained about half of its wetlands (110 million acres) between the 1780s and 1980s, mostly to expand farmland. Forested wetlands, like a good part of Reelfoot Lake WMA and the remainder of the Mississippi

Figure 34. Redwing blackbird on its spring singing perch; and male widow skimmer dragonfly.

River Floodplains, have shrunk about 85 percent since the 1960s.[1] State and national wetland inventories, however, can be nebulous since they do not differentiate between native wetlands, modified native wetlands, or manmade or artificial wetlands. That makes a big difference because there are far more inferior manmade or artificial wetlands than superior and essential native wetlands. Most native forested wetlands around Reelfoot were converted to agriculture by the 1960s. Along with this conversion, the lowly native "weeds" were eliminated.

"Weeds?" Yes, common every day weeds. Who would have imagined that? Well, it's easy when you think about bobwhite quail and other birds. Their abundance is directly related to the quality and abundance of native "weeds." Once there were plenty of these weed patches. That was way back when we farmed with mules or John Deere tractors with a flywheel to start the machine, the years before huge diesel farm equipment and "clean farming." Weed patches were in fence rows, along streams, in gardens, behind the barn, and so on. That is the habitat we are missing today—and the very thing quail, rabbits, and other birds desperately need. With the arrival of soybeans and clean farming, nearly all has been lost.

You might miss the manager's effort to provide for these different species of wildlife, unless you get a heads-up. Currently, this habitat is left up to public lands to provide. Notice the native vegetated buffer zones around Black Bayou Refuge, Reelfoot, and Lake Isom National Wildlife refuges— these vegetated buffers make up perhaps 90 percent of native plants and remnant wetlands in the Reelfoot region. Herein is another for managed wetlands like TWRA's Black Bayou Refuge. These manmade wetlands are usually developed species-specific for waterfowl, but we now know target species like waterfowl are not compromised by adding travel lanes and odd areas for other wildlife: furbearers, marsh birds, shorebirds, and half a dozen other species. In fact, this practice can enhance habitat for waterfowl. Wooded corridors and weed patches between fields provide disturbance security and food plots that provide acorns, wild millets, insects, and other food sources for waterfowl and all kinds of other wildlife.

Soybeans replaced hardwood forests. These wet areas were forests that flooded annually prior the 1960s. Hardwood forests produce acorns, pecans, seeds, sedges, and other favorite duck foods available to waterfowl (and a tremendous variety of other wildlife); when cleared, the entire forest ecosystem is sacrificed. Within a single decade, most of those patches of trees, remnants that produced seasonal wetlands, were crossed off the map—gone.

Figure 35. Mud snake, harmless but
vicious-looking.

Notice that some forest losses have been partly compensated by reforestation on the TWRA's Black Bayou Refuge and buffer zones recently
acquired and reforested around the lake. Fortunately, we have protected
forests in the federal's Long Point, Grassy Island, and Isom Lake refuges.
A few private hunting clubs delay dewatering, which provides spring habitat for waterfowl, shore and wading birds, and other wildlife after hunting season. Still, many marshes, basins, and other nooks, in and around
the lake, provide favorable wildlife habitat.

To be sure, the Reelfoot ecosystem becomes more valuable yet vulnerable every day. Questioning the worth of any natural or well-managed wetland, especially Reelfoot Lake, is folly. Like most living here, I once took for
granted that our great migration by some 1.5 million waterfowl were simply annual perks for living at the lake; I had not considered the millions of
non-game birds that annually visit the lake, let alone our native resident
species of flora and fauna. And who can imagine the contribution of the
lowly insect to the massive food web that feeds hungry birds, reptiles,

fish, many mammals, and more? And every insect is very specialized in the kind of plant it needs to survive—a low diversity of plants means a low diversity of insects, most of which are extremely beneficial—even essential—to a healthy ecosystem and humans. Where would the monarch butterfly be without milkweeds? Where would the prothonotary warbler be without caterpillars to feed their young? You can find the answers to these mysteries by procuring a copy of *Bringing Nature Home,* or any other book by wildlife ecologists like Dr. Douglas Tallamy.

The attraction of Reelfoot Lake to thousands of annual visitors alone should be enough to attest to the lake's great value. Those who live here and depend on it for a livelihood need not be convinced. Quality wetlands all along the great Mississippi River have been drastically replaced by one kind of development or another. Consequently, we find a corresponding decline in a dwindling population of migratory birds, fish, and wildlife, and other native flora and fauna; some have vanished entirely, not to be seen again. Will and good land-use planning will go a long way toward a reversal of that trend.

Chapter 11

Waterfowl Hunters, Duck Blinds, and Sportsmanship

Waterfowl Hunters and Hunting

Waterfowl hunting is a popular sport among Tennessee hunters. Annual state waterfowl licenses sales indicate there are between eighteen thousand and nineteen thousand hunters in the state, and most of the hunting is done in West Tennessee. Guess where the main hotspot is: Reelfoot Lake.

Nothing disorients a waterfowl hunter more than the anticipation of opening day of waterfowl season—careers and domestic life are at risk until the season ends about the first of February. Some defined the epidemic as "Waterfowl hunting fever": the walking state of a feverous syndrome that becomes rampant about the first killing frost.

Symptoms are something like malaria: anxiety, frigidness, and a low-grade fever prevails, undetected by a thermometer, and it returns periodically throughout the year. You see them hauling brush, rigging decoys, or building duck blinds in the middle of summer. It's probably camouflage being hauled to a duck blind, or it might be a floating duck blind being towed across the lake. You might notice boats that look like brush piles going across the lake. Most likely, you'll see this activity anytime during late summer and fall. The ailment can last a lifetime. All of this for a sixty-day hunting season.

Waterfowl is the big game of small game in the Mississippi River flatlands; or, insofar as the waterfowl hunter is concerned, the only game during late fall and winter. It causes considerable conflict for those who also chase little white balls over hill and dell, and club them to shreds. But give them due credit: waterfowl hunters may have more stock in wetland management than any other sport. Why? Of course, they pay licenses and fees that help pay for their sport. Tennessee waterfowl hunters spend nearly $2 million annually on state and federal licenses and permits alone.

Waterfowl hunters make another contribution: ducks need water 24/7, so waterfowl hunters have more appreciation for wetlands than any other category of users (only because others have not learned the enormous benefits to other wildlife). Fresh water is needed for ducks to nest, raise their young, rest, and make available sources of food; and water is needed to attract waterfowl for hunting. Sometimes conservation-minded waterfowl hunters maintain their duck ponds until spring—a good thing! Leftover duck-hunting ponds can benefit other wildlife: reptiles and amphibians, furbearers, predator birds, shorebirds, and numerous other wild creatures. Most waterfowl hunters are aware that winter wetlands like Reelfoot Lake are almost as important for waterfowl populations as prairie potholes, the habitat that produces most of the ducklings we expect to see during fall migrations.

Waterfowl hunting, however, compromises one of the most important features of Reelfoot Lake: tranquility. Tranquility speaks in nearly every scene observed, photographed or imagined at Reelfoot—except during duck season. So be prepared for a lot of noisy activity from opening day to at least the end of January. During that period, the "peace and quiet" we so expect of the great outdoors might be next to nil. Serenades of the most erasable sort begin long before opening duck season. The instrument is a duck call. Tune-up usually begins in the living room of the hunter's home. Check with the wife of a duck hunter, his family, or his friends; you'd think the hunter's loved ones might be on the duck hunter's side.

Wrong!

I was reminded of that dilemma early in life—the day I brought home my first duck call. Duck calls, as a work of art, are about the only time the instrument is admired by a non–duck hunter. Mine was one of those. An old, highly polished, walnut Bean Lake call I had was better crafted than any of my mother's furniture. I don't think I owned a hunting gun at the time, but I clearly remembered practicing with the duck call. Or maybe it was the offence it caused my mother and sisters. Not so politely, I was asked to leave and take the call with me—meaning evicted. The only way to make amends was go someplace beyond hearing distance to practice; another county was suggested. Universally, it seemed, "music" to a duck hunter is clearly an oxymoron to the general public. No instrument on earth of the wind kind is apparently more irascible to ordinary people. Not surprisingly, ducks generally seem to think the same thing.

All of this, however, can be dismissed at the annual Reelfoot Duck Calling Festival held at Samburg. Hunters, who make these calls, are truly craftsmen: a Johnny Marsh, an old Bean Lake, a Tom Turpin, a Larry Hickerson, or one of Terry Norris's beautiful duck calls are certainly works of art. And all of this can be greatly appreciated, until the calling begins.

I live along the lakeshore in the Gray's Camp community of nonresident cabins. A weekend or two before opening day of duck hunting, I often hear them tuning up, maybe as far as a quarter-mile down the way. Not intolerable before midnight. But go to the waterfowl festivals held at the lake and see for yourself; you'll find a bunch of enthusiastic and friendly folks with a lot of stories, waterfowl-hunting artwork, memorabilia, and gorgeous duck calls.

Yes, there is an art to duck hunting. Intense and red-faced from the strain of the wind instrument, some rare callers seem to be regular duck traffic controllers. Duck calls and duck calling have always been a big item here at the lake. You'd be most impressed to see a flock of fifty or so mallards circle down from the heights in response to their calls and their cautious approach to a spread of decoys. Wings cupped, the flock evaluates the landing zone carefully with every circle. It is quite a thrill to watch. The ballet might last ten or fifteen minutes before the finale at the landing—or they might get suspicious and abort the landing. Several local authors have written books on waterfowl hunting at the lake. One I can recall is *Reelfoot Lake Remembered* by Russell Caldwell. These books will take the reader to new heights in duck calling and hunting during those colorful years.

My grandfather Wiley "Pappy" Johnson was a champion caller, so I have been told. I've hear it many times; the record he made of his calling. Old-timers told me he needed only one live decoy and his trusted duck call to outdo all others in the vicinity to coax receptive flocks out of the clouds. A story about Pappy circulated around Walnut Log for years. Pappy, they said, was outdone one cold January day, when he called at a high flock of ducks for more than an hour with no results but frustration.

He had only one good eye (lost to a tree limb back in his lumbering days), so his depth perception was a bit off. Turned out, he was calling all the while to a small spider in a web only a few inches above his head. They accused him of having a nip. He never mentioned the incident.

Still, Pappy was apparently a well-known duck-hunting guide, and it was common in 1915 to get the daily limit of fifty ducks per day. There was no sport in Lake and Obion counties with higher esteem than duck hunting, back then as now, but it seemed a little too much to call these early hunters "sportsmen"; they killed and sold them by the boxcar loads.

Figure 36. Tommy Lovell's duck blind and sunrise mallards.

Salted down and dressed for the markets, ducks were preserved in stave barrels and sent to big eastern cities. It was more than the waterfowl population could stand. So it was a good thing the federal Migratory Bird Act of 1918 put a stop to that, or there likely would not have been a bird remaining in the Lower 48 but pen-raised chickens.

Sports Afield, or a field man from one of those popular sporting magazines, came and recorded Pappy's calling sometime back in his younger days, about the mid-1920s. He demonstrated his calls to me as a kid, and I listened to that old record after his passing many times on a hand-cranked Victrola. But, alas, when I searched for it some years later, I was told it burned up in a house fire. I tried to duplicate his techniques, but no one ever called me a master duck caller.

Back to the opening day of duck season and the hustle and bustle: the atmosphere is energized with hunter activity. Traffic around the lake begins an hour or two before official sunrise. Pick-up trucks and boat trailers may be followed by buddies in another vehicle, forming a caravan. They are all headed to boat ramps. Sound sleep is over for ordinary folks along forty miles of highway around the circumference of the lake. The gods of duck hunting forbid you living near a boat ramp. The murmur of voices, the whine of boat motors, and the anxious whimpering of lab retrievers speaks of excitement; it's a favorite time of the hunt, and they love it. The magic day begins about 3 a.m. on opening day of waterfowl season; it crescendos the first thirty seconds of legal shooting time. Not surprisingly, a fusillade from a thousand shotguns before sunrise awakens the entire marsh, not to mention sound-sleeping people who lived around the lake.

The prospect of peace and quiet is pretty much over until the wee hours of the evening, even though shooting hours end on the lake at 3 p.m., every day until the end of the season. I know that ilk of hunters—I've lived it. Absent motorboats, and all the modern fads, the art of waterfowl hunting at the lake is nearly as old as the lake itself. Market hunters, like my grandfather, could hardly be called sportsmen, although they loved the sport. But it was mostly an occupation—they made a winter living from the harvest of waterfowl. With the right load of No. 6 shot, the right gun, and the right hunter, one shell could collect enough feathers from twenty-five or thirty ducks, enough feathers to make pillows for a small family.

Gentlemen sportsmen followed the days of market hunting and elevated the sport to a new level, something I'd imagine was passed on from their European fathers. But Mr. Herb Parsons was probably the most famous sportsman to hunt Reelfoot Lake; he was not only a national shooting champion but also a national duck-calling champion. Born in Somerville, Tennessee, during the early 1900s, his hunting attire could not be mistaken for a poacher's; snappy hunting clothes with a neck tie were not uncommon. He helped set the standard for ethics and the new-age sportsmen, which shows in many of the old black-and-white photographs. They often used light-weight guns like .410 and 20 gauges using standard No. 6 shot shells. Even with heavy load shells, they considered a duck more than thirty yards away to be stretching a shot. My father told me not to shoot ducks responding to a neighboring hunter's decoys, no matter how close they came to our blind. We could not be called "gentlemen" hunters but we applied many of their standards.

The sport of waterfowl hunting continues to evolve to this day, depending on the definition of "sporting." A 10-gauge poly-choke gun set to infinity with buck shot and 3.5-inch shells fall outside my definition of "sporting." If that were not enough, some use custom shells to get an eighty-yard "sky-bustin'" shot, whether the ducks are in range or working their decoys or not. This brand of hunter has long since forgotten the term "sport hunting." So it is that many modern-day hunters are likely to

Figure 37. Elbert Spicer hunted ducks from
the tip of a cypress tree during the 1930s.

wink at these "old timer fairytales." Let it be; those were the early days
of hunter respect for his neighbors who also had a spirit of conservation
awareness. The science of wildlife management was brand new back
then, and hunter ethics were on the rise. We still need to be reminded, for
those were the best of times for quality sport of hunting.

As expected, you will see a lot of duck blinds on Reelfoot Lake, some
worth a successful hunt, some not. If you are not a hunter, you might wonder what the excitement of the sport is all about. Waterfowl hunting can
be extremely enjoyable and a cornerstone sport for outdoorsmen who
care about the future of natural resource conservationism; people who
will help assure the extraordinary benefits of the nation's native wetlands. But these benefits and ethics ebb away as surely as chemical pollu-
tion when waterfowl hunting becomes a crowded sport, short of conservation ethics. Here at Reelfoot, some waterfowl blinds can accommodate
a dozen (or more) hunters, but that is for "modern" hunters who have a
different concept of quality hunting than their forefathers. The original
customized duck blinds were rarely built to accommodate more than four
to six hunters. Friendship, sporting ethics, enjoyment, safety, and hunting satisfaction are dramatically compromised after that. It requires a

definition of "quality hunting." Here's one: quality hunting is taking wild game by the highest standards of conservation ethics. Quality hunting, in my opinion, takes a sudden nose dive when these characteristics are missing. Any justification in the guise of hunter opportunity after that is from ignorance of the sport or for someone's economic purpose, since gang duck hunting is no more sporting than gang hunting for spring wild turkeys; it's simply gang shooting, not quality or ethical sport hunting.

Actually, these camouflaged "lumps" called duck blinds are relatively new to the lake. So-called permanent duck blinds are blinds generally built by locals, many of whom made much of their winter income by guiding waterfowl hunters. These were very successful hunting blinds that had been hunted by the same person or families so many years that they were honored as that person's or that family's hunting blind. Ownership was rarely contested. While some of these blind sites are very old, the majority has been declared since the early 1950s. Before this, a common duck blind was more than likely a temporary, camouflaged "stump-jumper" boat. The boat was pulled up into giant cutgrass or button bush for cover and that often became the person's temporary hunting site for the day, or sometimes for the season. A charcoal bucket of sorts was used for heat and cooking. Where flat-bottom boats might be stranded on a stump, the stump-jumper's pointed bow was designed to skew off solid objects instead of upon them. The boat was also rather efficient at plowing through emergent vegetation, like giant cutgrass and "lily pads" (generic for lotus, white lily pads, and spatter dock). This hunting arrangement worked very well for decades as there was a great deal of respect and consideration for neighboring hunters.

The use of duck decoys has also evolved. My grandfather often spoke of using his half-dozen live domestic mallards as decoys. They were treated much the same as pets. These were very convincing to a flock of ducks. Too much, I assume, because they were eventually declared illegal to use. During my younger years, we used maybe two dozen artificial decoys (instead of hundreds used today) in the open holes of giant cutgrass fields and willows. This was "the spread." The only duck ethical to shoot was one that worked our spread of decoys, and that one had to come within thirty or forty yards of the blind to be within good shotgun range. Not that the limit was strictly adhered to, nor were we without competition, but a hunter worth his salt had to deal with guilt if he disrespected the unwritten ethics of the hunt. It was a kind of justified reciprocity without the law. And it worked well with a lot of wonderful hunting memories.

A new trend of duck hunting began on Upper Blue Basin about 1952, when James built his open-water duck blind. As a kid, I knew of no other duck blinds out in the open lake, let alone the "permanent" blinds you see today. Guys like "Slingshot Charlie" Taylor ("Reelfoot David"), long said to bag a few crippled ducks and small game with a slingshot, hunted best from a boat blind in the cutgrass. Guys like Elbert Spicer had a reputation for merely hunting out in the lake from the tops of stunted cypress. I've tried it but it is a tricky kind of hunt on a cold day, strapped to a limb while trying to sip a cup of coffee and shoot a duck. But no man-made "cabin duck blinds" were in my neck of the woods; only boat blinds camouflaged with switch cane and a little giant cutgrass were used.

That changed one season when James Scheland built a box-like, two-man "cabin blind" out in Upper Blue Basin, the only open water duck blind I knew on Reelfoot Lake. It was a gamble; no one thought a duck would come close to such an obtrusive structure. A cabin blind, as one might conclude, was something you could hide in, get protection from the elements, and almost live in. The top—half open/half covered—was designed to allow hunters to shoot from or duck under a hood to escape cold wind, rain, sleet, and snow (later, a boat blind was added). The entire monstrosity was camouflaged with whatever suited the site. James's cabin blind was covered with cypress boroughs by a lone cypress tree known as the "Flattop Cypress," about three hundred yards west of the Walnut Log Canal.

James's duck blind was a smashing success. He and my dad were probably the first to try their luck hunting from it. One bitter cold day a few days before Christmas, I was added to the guest list. Happily for them on previous hunts, huge flocks of mallards paid little attention to the obtrusive little structure. Even better, ducks working the decoys were unhampered by interference from other hunters since they were in the marsh, a quarter-mile away.

There was a problem with a two-man blind: who would hide the boat? To solve the problem, a guest was invited. Guess who that "guest" might be? Me, the third-grade kid who begged to go hunting. They explained right up front that my job was only to carry off and hide the boat, and to retrieve downed ducks. And they killed many.

I wasn't sure if duck hunting was worth freezing to death; but, driven by some internal drive to hunt, I could not resist. In more or less a windy lake blizzard, the heads of white caps were sheared off in a spray that created a smoky mist about six feet high across the lake. Not even the warmth of a charcoal bucket was offered. I just sat hunkered down in the far grassy shore some three hundred yards from their cozy duck blind. It would not be long before a flock settled over the decoy spread, which was often. By the time I retrieved the downed ducks and returned to my hiding place, the boat and I wore a thin layer of ice that shattered like fine snow or frost at the slightest wrinkle in my clothes. Before long, in an ice-coated boat, there I sat, sculptured by an unconcerned lake blizzard into a virtual snowman.

Thank goodness, about mid-morning the hunters allowed me a brief break. Was it gracious pity, or was it to stop me from daily bugging them about going hunting instead of going to school? It didn't matter. Teeth chattered as I tied the bowline and climbed stiffly into the warm blind. Heaven! Hot coffee from lake water and a biscuit with sizzling baloney and eggs were waiting. The boat, still tethered loosely to the blind, swayed back and forth, conspicuous as an out-of-control kite in the crisp north wind. Hardly had I taken a bite of my sandwich before a flock of some 150 or so mallards showed up high in the north.

Always scanning the sky, the skilled hunters were vigilant. James and Dad, out of reflex more than expectation, quickly sacked their sandwich and showered down on the flock with highball calls. They knew it was useless (any working duck would be shied by the boat, for sure). Nevertheless, as habit would allow, they had to give it a try.

Amazingly, the entire flock began to break ranks with cupped wings and descended like a squadron of acrobatic P-38s. It was probably their first pit stop since leaving Canada. The flock regrouped into a tight formation and circled—forever, it seemed. Struggling with a crick in my neck, I peeked through the cypress camouflage, when . . . whoosh! The sound of screeching wings suddenly filled the heavens high overhead. Wings set, they made their next circle. Now lower, the flock had made up its mind; they were on final approach!

James thoughtfully commanded that I pick up his little single-barrel .410 gauge, loaded and idly parked in the corner of the blind; it was good for one shot. Half of the biscuit sandwich was still in my mouth when I picked up the little gun for action. And there, out front, against the north wind over choppy water, the flock's wings hardly moved as they lost altitude in a wide approach to settle among the decoys.

What a brilliant sight it was! Twisting and maneuvering for position, the sky had turned into a living cloud of mallards, preened and brilliantly colored, fresh from their summer brood grounds. Over little more than three dozen decoys, the flock had been tricked. A more beautiful sight of nature is hard to imagine, so close. Feet stretched, orange-red in a blue sky, and as bright as Christmas lights, they hovered over the choppy water. Finally, they "pulled off their boots" and hovered twenty yards from the blind, half in the decoys, half still dancing above trying to find an unoccupied place to land.

Too late for all to settle, the command came: "OK, get'em!"

It was over in a flash. Six mallards lay kicking belly up on the water when the volley ceased. I claimed one. The savvy hunters smiled and offered me cheerful congratulations. "You've just helped make Christmas dinner, my boy!" Dad said. Well, how could I miss with so many ducks?

Word spread, and by the next season new "open-water duck blinds," boat blinds and all, began to show up across the lake.

Ethics and the use of decoys changed as more and more "permanent" blinds were declared. By the 1970s or 80s, old traditions and ethical duck hunting were but a glint in the old timer's eye; a new age of competitive hunting had arrived. The flack and lower quality hunting seemed to be directly related to the overuse of "permanent" duck blinds on public land. The trend reached its peak in the mid-1990s. The spirit of duck hunting has always high among hunters, but when "permanent" duck blinds, unregulated, became a fad, competition and stress of the sport became unpleasant for ethical and quality hunting Noticeably, there was more activity, for one thing. Not that it was undesirable, but the sport became a year-round ritual. Building and repairing duck blinds and pampering the layout of "duck holes" only added to the sport, certainly a good thing so long as hunting ethics were part of the trend.

Nevertheless, it didn't turn out that way; heated disputes became more and more commonplace. The outcome too often brought on scorched duck blinds and the odor of diesel fuel. "Indestructible" metal blinds were built, but these were caught up in the fray as well; TNT made piles of scrap from some of them. Too many duck blinds emphasized competition; competition caused more disgruntled hunters; disgruntled hunters sometimes rained down pellets on the heads of competitors; and honest-to-goodness hunters who respected the sport hunters wanted relief. As of this writing, I don't recall any serious physical altercations more than a few bruises between these hunters. But the potential was there and something had to be done about it. That happened several years ago when the state regulated the distribution, occupancy, and construction of duck blinds. A new day for duck hunting had begun. While designated blind sites are still too close in many cases for quality hunting, relative peace has since prevailed.

Figure 38. Tennessee wildlife agents: Removing illegal duck blinds
to restore order and equitable hunting methods.

Duck blind drawings were like a hunter holiday. Conducted by TWRA the first Saturday of every August, it was a very enjoyable event with a lot of excitement. One would think the gathering was a raffle for buckets of thousand dollar bills. Three hundred or so wishful hunters showed up each year for the chance to draw one of some sixty or so public duck blinds.

The remainders are "grandfathered" duck blind sites, those where tradition was conceded; chosen sites for those who had hunted in them for decades could register and keep them. All of the 193 registered blind sites on the lake are destined to become open to the public for the annual drawing, once the grandfathered blinds become obsolete. That happens when attrition by failure to register or death removes the original owners and their party or other regulations apply. Waterfowl hunting is a very involved sport at Reelfoot, and they are acutely aware that without healthy wetlands like Reelfoot Lake, the sport of waterfowl hunting would be finished.

Chapter 12

Exploring Reelfoot Lake with the Seasons of the Year

Spring

Early Spring (Mid-March—Mid-April)

Freshet spring showers have raised the lake and the excitement of many creatures by early spring. Low ground, like that on much of Black Bayou Refuge, has a few inches of water on it. Spring peepers waste no time setting up choirs for their special archipelagos, exquisite musicals with bell-like tones—a signal that winter is over. It usually happens in mid-March. So, if you are passionate about the outdoors, don't miss this introduction to spring.

Sport fishing and sightseeing are the most popular outdoor activities during this season. Look for low ground and small pools of fresh water—like some you'll find on Black Bayou Refuge. Get out and look them over because these pools have the table set for birds and mammals whose foods are crawfish, minnows, insects, frogs, and reptiles—all of which are attracted to this kind of habitat.

Spring activity flourishes with wildlife. Baby turtles have been flushed out of their nests from the previous spring, and are likely to be crawling or sunbathing on the roads. So watch to avoid them. The first ospreys arrive to claim nesting sites, and migrant bald eagles begin to leave for more northern climes, but resident eagles stay and can be seen carrying sticks to refurbish old nests. Birders already know about this, and outdoor photographers have waited all winter in anticipation of it. Hunters, exhausted from a long season, can be all of these, believe it or not. But some folks are simply glad winter is over; one tweedy bird is like another, but the warm breeze and the freshness of nature all around is somehow medicine for "televisionitis" and cabin fever. No matter, if you think you might like our style, slip on a pair of knee boots, load up a canoe, and break away for this special season. Reelfoot Lake is the place you need to be.

Nothing satisfies the need to be outdoors more than a hike on a well-maintained hiking trail or a float on a nice boat trail. Besides poking around the lake by way of an auto tour, a hike on boardwalks and trails is a refreshing treat. One of the best, however, is to take a quiet watercraft trip along the canals, boat trails, and bayous. These waterways are particularly good alternatives on windy days. The lake begins to whitecap when the wind reaches about ten miles per hour. Do not be tempted to take a shortcut and cross open water in light watercrafts under these conditions. Good launch sites to avoid or minimize wind effects are at Samburg, the Kirby Pocket area, Walnut Log, Grassy Island, the Air Park Campground area, and Gray's Camp.

Reelfoot has many canals and boat paths. Most major canals were dug with a floating dragline between 1943 and 1951. But, today, you might see two major types of machines in use to keep canals and waterway navigable: a Cookie Cutter and a Jet Spray. TWRA and Reelfoot NWR both participate in the maintenance of the lake's canals and waterways. The Cookie Cutter looks like a monster out of science fiction, but it is a pontoon-like barge working with rotating, churning, fan-like cutting blades on the front; the machine is used in zones of thick aquatic vegetation. Cutting and shredding only vegetation, it leaves a boat trail no wider that the cutter blades. The machine, first on the lake during the mid-1950s, is owned by Reelfoot National Refuge and operated by both wildlife agencies to clear aquatic vegetation that often clogs boat trails and channels.

The *Jet Spray,* owned and operated by TWRA, might bring to mind something like a velociraptor from Jurassic Park. The machine is stationed on a large barge with a hydraulic vacuum arm and rotary cutting ball to loosen bottom sediments. Then, a vacuum pump jettisons the loosened muck and debris overbank in thin but harmless layers. The machine is used to deepen navigation channels. Channels are usually excavated twenty-five feet wide and six feet deep. The operation is environmentally friendly and very effective in keeping these canals navigable. Over time, major canals with mysterious names like Horse Island, Donaldson, Green Island Cut-off, and Bayou du Chein fill in after some years with sediments, and the jet spray is the ticket to relieving that problem.

De-snagging is a supplemental operation by TWRA to keep the channels navigable. It is not unusual to find agency technicians with winches and chainsaws involved in removing large logs or fallen trees from the channels. In 1986, a major effort was conducted with a TWRA crew to remove large debris and logs, many sunk deeply in muck. This work made a huge difference in the navigability of these channels in motor boats.

Canals can help you stay out of the wind on breezy days, retreat from an unexpected storm, and be a photographer or birder's best friend. If the going gets too rough, with luck, you might find a stout duck blind for safe cover, but check for wasp nests before you enter. Boat traffic during this season is likely to be light, since all duck hunters are worn out and need a break before tending to duck blinds and decoys.

But a few bass and crappie fishermen might pass by, figuring you are lost but will eventually find a way home. Pay them no mind; one day they likely will join you. But these man-made channels for navigation are also scenic and excellent for observing or photographing wildlife. Animals such as muskrats, raccoons, and opossums can't resist bathing in warm sunny spots along the banks of the canals, although I have seen energetic mink do the same thing. Often, still sleepy from winter dens, some are so tranquil they might yawn and refuse to move when you pass. It is a time when buds everywhere are swollen fit to bust, and the red leaflets of water maples are about the size of a mouse's ear; rejoice—the time is perfect for spring adventures.

The woods are usually quiet during these early spring adventures. But don't be surprised if, overhead, the unpredictable, nerve-wracking

Figure 41. Marsh birds on Reelfoot Lake WMA:
Black-neck stilt and least bittern.

screams of the wood's spring prima donna causes your hair to stand on end—it is only the "Good god!" Most know it as a pileated woodpecker. Such excitement is often settled by the eight-note call of a barred owl. Then, deep within the cypress swamps, you might hear the silence broken by what seems a chaotic gathering for mayhem. Not to worry—if it sounds like the squawks of desperation, it could very well be a heron and egret rookery, often fifty to a hundred feet high in the tall cypress groves.

Visiting a rookery can be quite an adventure many places at Reelfoot, but if you are not a seasoned woodsman, a park ranger, who sometimes schedules these trips, could be your best bet. One such trip by a park naturalist is called the Deep Swamp Canoe Float, which can be a very enjoyable outing on its own. For those who have a special interest in these rookeries, several can be located nearby on islands of the Mississippi River.

When you see all of this, spring migration is certainly imminent; that is, vacationing winter birds will be on their way back north to the excitement of nesting and raising young. Keep a sharp eye. If you are curious about birds, some 325 different kinds are known to stop-over here on their northerly quest. Some, like the bald eagle and osprey, will stay to nest at Reelfoot. Many of the two hundred or so neotropical birds can be found at the lake, such as tanagers, wood thrushes, purple martins, vireos, warblers, and shorebirds—they migrate from their summer breeding grounds of North America to their wintering grounds in the tropics of Mexico and Central and South America. One you can't miss on these routes is the bright yellow prothonotary warbler. A canary-size bird (sometimes even called a "canary"), it loves bayous and canals and often seems to follow a boat, hopping from bush to log in search of caterpillars for its young as it goes. If you see this lovely bird, it probably just returned from South America. Your birding journey is just beginning.

Although most birds are hatched and raised considerably north of their wintering grounds, many (like herons, osprey, white pelicans, blackbirds, and several others) nest in wide latitudes. Some birds we see here, for example, have kin that spend their entire lives along the southern coast and Florida. But Reelfoot Lake is the crown jewel of wetlands along the Mississippi, an oasis so critically important to these birds that it is difficult to imagine how vulnerable migrants along this flyway would be without it. Sometime during the year, Reelfoot will harbor nearly every one of these birds, if just for a little while, to refresh so they can continue their journey to another destination.

Summer residents such as bald eagles, osprey, least bittern, moorhens, herons and egrets, numerous warblers, redwing blackbirds, spotted sand-

Figure 44. Least bittern.

pipers, avocets, lesser yellow legs, and others are fairly common. A little review will help you distinguish the residents from the migrants. Tiny warblers, besides the prothonotary, are abundant during both fall and spring migrations; they are hardly noticed except by dedicated birders. One reason is that the colorful warblers usually stay high in the crowns of the forest. Looking up into cypress trees that grow more than a hundred feet high can stretch our means to see them. That's why I mention that these shy birds will come all the way down to the ground to take a bath in a natural or well-designed stream of water. The stream must tumble to resemble the sound of a small, babbling brook.

Figure 45. Author's fresh catch:
Two-pound crappie.

Birds can hear even the slightest sound of a stream from a long way off—a good feature for managers to consider in areas like the Ellington Center. The stream need not take up much space so long as it is well-placed among mature trees. It should have perhaps three or more levels of natural canopy as vertical stairways for the birds, and with adequate cover around the stream so the birds feel secure. These magic nooks are not easy to find at Reelfoot because of the flat terrain, but they can be fairly easy to construct.

For fishermen, early spring is the prime season for largemouth bass and black crappie fishing. The same is true throughout mild winters, which can begin in February, which is a good time to fish the shallow sub-basins, up next to the tree roots and bushes, in the shallow water two feet deep or less. This is prime time for these fish to spawn. Long poles and short lines with jigs and wax worms or minnows are good rigs. If the bottom of the lake in the shallow areas has experienced summer drying often enough, the bottom will be firm around stumps, the roots of cypress trees, and bushes with many minnows and insect larva, excellent places for spawning fish. So gear up and enjoy the spring fishing.

Late Spring (Late April—Mid-June)

Fresh fields of aquatics with colorful flowers suddenly emerge in spring. Lotus with a large yellow flower; spatterdock with golf ball–sized yellow flowers; arrowhead with a spike of white flowers; the fragrant water lily with its fragile but brilliant white bloom; pickerel weed with its beautiful purple flora, and other aquatic flowering plants—all emerge during spring. You'll find them in shallow water of about five feet or less along the shoreline or in small open basins. The flourish of spring is the busiest time of the year for wildlife and people. Auto tours are generally a good option about any season of the year, especially when the weather is a little "iffy" for those not up to more arduous activities. Otherwise, hiking trails, or watercraft trips through the canals, bayous, and ditches can be very good options. Spring is usually a season of dependable weather and a wonderful time to be out on the lake. The State Park begins pontoon trips this season—very enjoyable and neat for family or a small group of friends.

Aquatics in the early stages of early growth are easy to navigate until about the end of June, when plants like spatterdock, American yellow water lily, and pickerel weed emerge above water and become taller and denser. They often make good fishing areas for bluegills and crappie since the water is usually cooler with high free oxygen content. These aquatics usually give way to giant cutgrass fields as the lake becomes shallower to-

Figure 46. Pileated woodpecker
at Horse Island Canal.

ward the shoreline. Be cautious when dealing with this grass since giant cutgrass can slice human skin like a razor, but only if you press against it with a downward motion. The dense growth of aquatics during middle and late summer is no reason to avoid the lake. There are still plenty of open boat lanes, canals, bayous, and open lake to maneuver.

Occupying the same space as cutgrass is a plant called "water willow"—an exotic. You might wonder about the nature of this woody invader, as it is just about everywhere on the lake: water willow is truly "a curse to Reelfoot Lake." The exotic is a vine-like, woody plant with small leaflets that grow over and smother native aquatics, including giant cutgrass. Other than cover, at Reelfoot, I know of no benefit whatsoever of water willow to fish or wildlife. On the contrary, it eliminates the nesting and food sources provided by giant cutgrass. I saw only a few sprigs of this plant growing on logs when I was a kid living here. It appeared innocuous at the time, but today it has choked out nearly all of the old cutgrass fields on the lake. Navigating through it is problematic—even an air boat can't go through or over it. Realizing its detriment to outdoor recreation and the ecosystem, TWRA has reluctantly resorted to EPA-approved herbicides to tame this invasion. The effects of this control are noticeable throughout the lake, except in the federal refuge; so far, the agency has not been cleared through federal policy to use this treatment.

Figure 47. Invasive water willow (leafy-green with dead spikes) encroaches on and smothers the last of native giant cutgrass at the water's edge in Brewers Bar.

"The curse of Reelfoot Lake"—once the name applied to giant cut-grass—is reputed as an invader of shallow basins—and that's a fact. But it was labeled a "curse" because it was a hindrance to navigation. This was a major reason canals were dug—as access for boats through dense aquatic vegetation. But the reputation of giant cutgrass should be reconsidered because it is native, and the invasion is a part of natural succession. As wetland fills with sediments and progresses toward maturity, shallow water promotes emergent plants like giant cutgrass. It is all part of the natural process. Cutgrass will eventually be followed by cypress trees and other woody plants.

But giant cutgrass is also beneficial to wildlife. It's a good nesting habitat for all kinds of birds (redwing blackbirds, least and American bitterns, puddle ducks, and others). You will often hear the familiar "*coo-coo-ca-coo*" of least bitterns, or the "*o-ka-ree, o-ka-ree*" of redwings as they cling to a blade of cutgrass singing their hearts out. Wood duck broods seek cover in and around giant cutgrass. Muskrats relish the tender stems of cutgrass, and it provides protective cover for dozens of wetland mammals and reptiles. The seed head has small seeds that songbirds and ducks eat. So we should rethink the notion of giant cutgrass being "a curse."

Osprey and bald eagles were both rare birds at the lake when I was a kid. Both, once endangered species, are now common. Ospreys have been called "the best fishermen on the lake" for good reason; their diet depends heavily on fish. They might appear to be eagles because of their size and affinity for open water, but they are classified as members of the hawk family. I don't remember seeing an osprey on the lake as a kid. Imagine, in 1978, only two pairs of osprey were known to nest in the state, one pair at Reelfoot Lake. Two young were hatched in 1981 and three pairs of fledglings in 1985. Osprey and bald eagle nests are both large and usually in trees. But osprey nests are usually about half the size of a bald eagle nest. Non-game biologist Don Miller put up several platforms in the open lake to encourage visiting osprey to stay and nest. You might see some of these old structures today. It worked well in Florida and other places, but at Reelfoot, osprey seemed to prefer low-growing cypress out in the open lake. I have counted as many as eight osprey nests in Upper Blue Basin alone, and today nests are commonly found around the lake.

Osprey chicks are still small in June and instantly hunker their heads down when a parent issues warning of an intruder. Their fuzzy heads might not be seen until one of the adults brings food, usually fish but sometimes a turtle or a snake. While these birds are well-acquainted with passing fishermen, they still object to encroachments. They screech, scream, and squawk, among other warnings, when boats get near. Stay a good distance from the nest so as not to unduly harass the dutiful parents. Get too close, and food will not be delivered, nor will you see the fuzzy heads of the young. But ospreys are familiar with people and, so far, I've not seen a nest abandoned because of human presence. The Reelfoot State Park has long conducted the annual bald eagle festival but the park and local tourist council jointly conducted the first Reelfoot Lake Osprey Festival in July 2021. There is a special reason why the timing of the event should be very popular—the young should be prepared to leave their nests.

With the rare sighting of a golden eagle, bald eagles are the only eagles found at Reelfoot Lake. Bald eagles became our national emblem June 20, 1782. Bald eagle populations, along with osprey and other birds of prey, were nearly eliminated during the 1950s through the 1970s. The cause was DD-T, an insecticide that interfered with the bird's calcium metabolism and caused egg shells to be weak and crushed by the weight of the parent during nesting. Eagle nests were absent here from 1961 through the 1980s. TWRA's non-game coordinator, Bob Hatcher, and a cadre of agency biologists and climbers made a concerted effort to change that. Between 1981 and 1988, forty-three young birds were released at Reelfoot by TWRA with assistance from State Parks (a practice called hacking). Today, there are about a dozen active nests at the lake and as many as two hundred wintering bald eagles.

Bald eagles, like osprey, feed mostly on fish, but ducks, coots, snakes, and other small animals will do. One was ambitious enough to take our pet cat from the dock about two years ago. We did not take kindly to the eagle for doing that, but had to look at it from the eagle's point of view as well.

Nature is sometimes unpredictable. Early one morning a year or two ago, Kathy and I were startled by a loud thump on our roof. The weather was calm and bright, and we saw no reason for the disturbance. Immediately, we jumped up and ran outside to investigate. Still fresh but not in good condition was a good size red-eared turtle lying on the ground, which appeared to have tumbled off the roof. Kathy concluded we now had "flying turtles." But overhead, a crow soared in tight circles, and a young bald eagle in a nearby tree watched what we were doing with the turtle. The crow could not have picked up the turtle but the eagle certainly could. There could be only one conclusion: the crow, jealous as they are, had harassed the eagle carrying the turtle. I concluded that the eagle had lost its grip and dropped the turtle on our roof. I've never known of an eagle to prey on turtles, but now I no longer doubt they do. And I have yet to confirm a flying turtle.

Bald eagle nests are more difficult to find than osprey nests. Eagles may not use the same nest as the season before, and nests are often remote and scattered. Nevertheless, don't be surprised to find one in a tree next to

Figure 48. This noisy juvenile bald eagle
entertained us the summer of 2018
begging its parents for food.

a busy resort. Young bald eagles have hatched by this time of the year, and they could well be standing on a limb near their nests begging for food, or getting flight instructions from their parents. Already with a wing span of six to eight feet, it will be another four to six years before the eaglet will be adorned with the white head and tail of its matured parents. Check with State Park rangers for advice on where to observe nesting eagles.

Bald eagles put up less fuss about humans than osprey, which may be because their nests are usually higher in the trees than the ospreys', and their dozen or so nests are often in isolated areas—but not always. If you are lucky, you might witness the spectacular midair courting ritual by a pair of these magnificent birds; they lock feet upside down, high above the ground, and spiral toward for some distance before recovering flight. They mate for life, but usually re-pair should a mate be lost. Last summer, we had a nesting pair of bald eagles that commonly perched in the tall cypress only thirty yards from our porch. We never located their nest but feel sure it was somewhere in the large cypress trees along our shore, since we could see or hear them nearly every hour of the day. Probably, it was the same pair that had to abandon a well-maintained nest in a large cypress near the Blue Basin Cove boatshed. Later we learned a female eagle with a broken wing had been found nearby. The injured bird was taken to the rehab center by State Park rangers. The eagle was healed and later released, but we never knew if the pair reunited. The parents of the noisy young bird in the above picture could well have been these adults.

Tree swallows are insectivores feeders found nesting here all summer. By mid-September they begin to colonize and show up in large numbers.

Figure 50. Tree swallows in a small cypress over water.

Figure 51. Airpark campground
trail photos.

Low-growing cypress scattered in the open lake are often filled to capacity with these very active and neat little birds. They have dark to irides-
cent green on their backs and white undersides. Often a tree-full will flush
as if on signal, and then fill the air like leaves or butterflies in a whirlwind.
Tree swallows spend the day socializing and picking flying insects, and
hopefully a lot of mosquitoes, out of the air with the skill of bats.

Large rafts of white pelicans are relatively new to Reelfoot Lake, although they have been recorded here. Like eagles, I saw none of them when
I was a kid. We saw the first large flocks on Upper Blue about eighteen
years ago, when about three thousand birds paraded past our dock like
proud peacocks herding shad into a small slough by the house. I haven't
seen them perform such a frolic since. The water thrashed so much that
little rainbows were created. The big birds still show up in large numbers
during both spring and fall.

Soggy winters and early springs can be a little too unpleasant for the
timid but, even then, watch for nice days and have knee boots or hiking
shoes, binoculars, and camera ready. These can be most rewarding times
to be out. Late summer through fall are easy times to be on the trails and
boardwalks. Hot days can be beat by getting out early and late. Southern
shore boardwalks and Grassy Island at Walnut Log are top-notch. Grassy
Island and Airpark Campground have excellent hiking trails.

Summer

Early Summer (Mid-June–Mid-July)

The freshness of spring begins to pass by mid-June. There are still very interesting things going on that many miss because they avoid the lake after the excitement of the spring flush. Sport fishing, sightseeing, and watching wildlife make up 67.5% of visitor activity on the lake during summer. Most days are pleasant all day, but mid-day can be in the nineties as summer progresses. Being on the lake during the coolest parts of the day can be very rewarding, and one of those rewards is the quiet solitude of the season.

By this time, bluegills are ending their major spawns and crappies have retreated to locations unknown. Still, some of my best bluegill fishing has been in late June. One might find bluegill spawning most of the summer, albeit with less enthusiasm than earlier spawns. Fishermen turn to other things—choosing not to face hot days, and searching for fewer bluegill beds and hidden crappie. Some years, however, sport fishing is poor for reasons often unknown. Bluegill fishing was the number one sportfish at Reelfoot when I was a kid. Bass fishing was popular, but a rare few fished for crappie. My dad said fishing for them was like catching "a wet sock." But he was a bluegill fisherman, addicted to the zing of line cutting water when a half-pound bluegill took a cockroach or "horseweed worm," which were favorite baits at the time. There seemed to be no end to the great size, abundance, and sporting nature of these scrappy fish. Crappie, popular on the menus of local restaurants, was legal for commercial harvest and off and on since the 1940s. But the commercial harvest and sale of crappie was made illegal, as mentioned earlier, in June 2001.

Bird watching is most popular during spring and fall—the seasons of major bird migrations. But it remains a year round activity. The spring rush of migrants is pretty much over by mid-May, and many have settled into nesting and brood-rearing habitat or gone farther north. So you will easily find herons, egrets, coots, prothonotary warblers, redwing blackbirds, osprey, and other birds of prey. Less conspicuous is the small head-bobbing spotted sandpiper. The first thing people as is, "What's that little bird?" This sandpiper, with a bunch of round dark spots on its breast, seems to be around well into summer. Its territory is along shorelines. Busy as a bee, you'll see it flitting stiff-winged from log to log searching for insects. It's about the only sandpiper you'll see out on the lake; the rest (nearly a dozen) are found in early spring—around wet fields and shallow water pools. Nearly all raise their tiny chicks above the Arctic Circle.

Figure 52. Forester terns, Upper Blue Basin.

Out in the open basins, you will see several kinds of white birds, mainly gulls. But one of the most elegant birds is the forester tern and the endangered least tern. Both of these birds have white bodies and black caps. The forester tern is about the size of most gulls; the least tern, about half that size. Both are inclined to settle on logs out in the lake. Sometimes several set up on a single log, and they are fairly tolerant of a boat, providing it does not go straight toward them. A favorite technique with photographers is to drift downwind slightly off course from them. Try it; you'll see.

Bull frogs are rarely heard spawning anymore, and I mention them for this reason. Something sinister has happened to them, and Tennessee biologists to my knowledge have not investigated the problem, although it could be a global issue. It is not unusual to find lethal viruses and fungi in reptiles, such as the recent turtle virus in the Everglades. All reptiles, it seems, have mysteriously declined in numbers for reasons unknown to me. Still, snakes of several species, green tree frogs and other frogs, and turtles of most species are quite common at the lake.

You know by now that osprey and bald eagles are likely the only two eagle-size accipiter you are likely to find nesting on the lake. The young this time of the year have grown considerably, and are half the size of their parents, hungry, and—unlike their behavior during early spring—not so shy about showing themselves. It is very rewarding to discretely observe the family conducting their lives. Momma and poppa, however, are as grouchy as ever about human intrusions, and they will let you know when you get too close. Be very considerate of their needs.

Figure 53. Bull frogs ready to spawn.

As mentioned earlier, bald eagle nests are often difficult to locate, even when you know you are near. They are not usually as fussy as osprey, however, while tending to their young, mainly because their nests are concealed and high in the canopy of trees. This year was a special eagle-watch year for us. We watched a young bald eagle grow up to flight age from our back porch. It was early June that a young eagle showed up begging for food near my dock, complaining that it was not getting enough attention from its parents. Chances are, however, the youngster had lost a sibling, since there are usually at least two young from a nest. Its favorite perch was the broken top of a cypress tree at our dock, only twenty feet off the ground. For at least a month, the young eagle flew back and forth across our yard between perches, begging its parents for food. They often seemed to tempt the youngster by carrying the food a little farther than it normally flew. It had hardly any fear, often flying no more than fifteen feet above my head, switching from one perch to the other. We were entertained all summer by these birds.

Turtles can be entertaining wildlife, too. Nesting turtles are common during late spring and summer. Watching them go through their nesting ritual is educational and amusing. Off hand, I can think of a half-dozen species that lay eggs in our yard. Activity is greatest for nesting turtles after a soaking rain shower; it softens the clay soil (gumbo: hard as a rock when dry and sticky as glue when wet) where they will likely lay their eggs. This is a happy time for most lady turtles because, otherwise, she must produce her own water to soften the soil for digging. We've seen as many as four or five turtles come rushing from the lake at the mere threat of rain; even the first sound of thunder usually will do it. Digging in the dry ground takes so much longer than after a soaking rain, and they well know it.

Lawns are perhaps the safest and most desirable places for turtle nests, since these places are not likely to flood or be plowed up by farming equipment. This circumstance is a considerable change since I was a kid. Back then, lawns were not so manicured with exotic, insect-resistant, lawn grass, and farms were smaller and numerous—a thousand-acre field might accommodate ten families, each with a hundred acres or so to live on. Those farms provided nearly a hundred percent more habitat diversity for wildlife, due to nothing more than vegetated fence rows. In addition, farmers often left vegetated buffers in odd areas and around crop fields, mainly because it was not worth cultivating. But some modern farmers also can have a recognizable streak of land ethics and leave field corners and fence rows for wind breaks, beneficial insects, and wildlife. The practice is extremely valuable to wildlife; it is not only suitable turtle nesting habitat but the last link in the survival of myriad wildlife. Today five thousand acres owned by a single landowner, without a single fence row, a patch of woodland, or even a blackberries patch is not unusual. Where will the turtle find a nesting place?

Turtles generally will not nest in the shade of a tree. Ironically, that's certainly a minimum loss for turtle nesting in Lake County. You might notice that trees are pretty scarce in local river floodplains. All farmers I know in the Mississippi River floodplain today seem to despise trees and make every effort to be rid of any standing. Little or nothing remains that might be suitable for field lark to perch, a rabbit to hide, or a bobwhite to nest, or even a square foot of fallow land for a nesting turtle. Even ditch banks are plowed to eliminate possible competition from the seed of a weed. The economic margin of the industry is so slim, I'm told, farmers risk knocking down stop signs, telephone lines, power poles, and mailboxes on road rights-of-way for an extra bushel of soybeans.

My source concedes that farming might be more profitable if one avoids altogether small troublesome places that risk damage to costly equipment and provide questionable profits. But the urge to farm every square foot of land or go bust is apparently part of the modern era of clean farming. And it is easy to see on floodplain farms, such as those in Lake County. Nearly every square foot of land from road to road and ditch to ditch (a fence is a rare thing in a floodplain) is bare ground after crops are harvested; on the other hand, the common protocol for progressive farming today might simply be misinterpreted.

Yes, farming practices are about turtles, too. Perhaps the farming industry should take another look at over-worked, too-wet-to-farm farmland, land often forced to produce profitable crops it has no capacity to accommodate. And it would seem most profitable to avoid high-risk nooks mentioned above—places where equipment damage and fuel costs do not justify the effort.

Surely I'm misinformed because I see it repeated year after year. At no loss to farm profits, it would seem better to spare these places and leave a bit of fallow field or grassland with native varieties of wild plants needed by wildlife, especially in places like corners, ditch and stream banks, road rights-of-way, and other odd areas for creature habitat—say, a place for at least a rabbit, or for a turtle to nest. The old phrase for it is "good farmland stewardship." After all, the private landowner, who has most of the land and holds the key to the future sustainability of fertile ground, fish and wildlife, and country, actually has the nostalgic characteristics of countryside and sustainable farmland.

It makes little sense to squeeze in these odd and difficult areas to farm when farmers on good crop ground and a good year stand to make sixty bushels of soybeans per acre. At a top price of $10/bushel, that would gross $600/acre. Sounds impressive, but my source says the net on his farm is something like 10 percent of the gross. If that is true, a hard-working farmer scratches out only six bushels ($60) of soybeans/acre net profit. It seems awful risky business to spend heavy equipment time to drain wetland on the chance that the seasons will be dry enough to be profitable. Why not stay out of the low-risk ground and plant trees, or leave as plots for wildlife, beneficial insects, and native "weeds" or forest? That said about turtle survival, there are other things to mention about turtles.

Lady turtles have strong, short toenails, stout and suitable for digging through a small opening (just large enough to insert a rear leg) to excavate a softball-size cavity in the ground, where she deposits her eggs (depending on the species: softshell and snapping turtles have twice as many eggs and twice the nest size). Males don't help, maybe because their toenails are too long (he uses them to grasp the shells of females during breeding). After twenty minutes or so, the newly constructed nest is finished. But she is very shy, and if someone or something should spook her, she will likely abandon the nest. Otherwise, the lady will lay a dozen or fewer clean, white, spherical, soft-shelled eggs into the nest and carefully seal it with a plaster of mud. Job done—she leaves the nest to an uncertain future. If good fortune prevails, the eggs will hatch by fall; the young turtles will stay put in the security of the nest until spring rains soften the ground enough for them to be freed.

But the turtle's life is rife with hazards. Here's the usual setting during nesting: a common crow (or a murder of them) is likely to be watching her every move. Even before the lady turtle begins to seal her nest, the crafty crow flies down to harass her, hoping she will abandon the new nest to make the raider's job of stealing eggs easier. Reluctantly, she usually acquiesces. It matters little; crows will dig up a nest, even if she seals it, and the eggs are quickly gobbled up (sometimes sharing, sometimes scuffling with his buddies), or the extra transported to feed youngsters.

So what if the crafty crow is elsewhere harassing someone else and does not know the lady turtle has laid her eggs? Is the nest safe? Not necessarily. Even before good dark, other neighbors with hefty appetites and sensitive olfactory receptors—well aware that turtles are on their nesting missions—are on the prowl. Raccoons, to name one raider, are as crafty as the crow. Raccoons surely have noses as sensitive as bloodhounds. Of the fresh turtle nests I have watched laid and sealed, nearly all that I recall are raided by daylight the next morning.

Armadillos, skunks, and opossums will also dig up turtle nests, or eat the scraps left by other invaders. Fishermen have used turtle eggs to bait trot lines, and some have used turtle eggs in lieu of chicken eggs to make cornbread. The turtle provides a great service to these lovable raiders, but it leaves one to wonder, How does a turtle manage to survive with such dependence on them by others?

Deep Summer: Take a Break!

After the fourth of July, hardly a soul is found on the lake. No reason to be idle adapting to the A/C, although it might seem better to sit on the screened porch with a glass of ice tea with a fan blowing. It is also tempting to wait until the day cools a bit and strike up the backyard grill for burgers. Yet, another option is to skip all the prep and go to one of Reelfoot's famous restaurants for a traditional Reelfoot Lake fish dinner.

Here are some other options. Take a break any time of the day and to go to State Park's R. C. Donaldson Memorial Museum, attend one of their fine programs, visit their nature center, take a sundown hike on their neat boardwalk, or check the park schedule for a pontoon cruise.

Quiet cruises or floats on the lake are the most underappreciated and underused activities at the lake, only because we underestimate the beauty and pleasures of it. With little effort and little expense, these trips can be one of the most rewarding experiences the outdoors offers. Sure, days are steamy, and we are a little low on energy, but the time spent to prepare and do it is expendable, hardly worth mentioning.

Only a few minutes of sleep is sacrificed to be on the lake by sunrise, so grab your gear, a snack for the road, a sip of something, and go. Have no other objective than to enjoy peace of mind, the gorgeous sunrise, and the early morning stir of nature. Company is great, but make sure they want to go for reasons as good as yours and know how to talk in a low voice; it's a requirement to enjoy the best of nature.

Photographers know to be camera ready from the first glimmer of sunrise until mid-morning. Don't concern yourself with insect pests. I rarely encounter a mosquito over open water; woolies or chironomids are often mistaken for mosquitoes. They are about the size of mosquitoes, sound like mosquitoes, act like mosquitoes, look almost like mosquitoes, and they buzz around a light—but woolies are harmless bird and fish food (a fit in the food web more than you'd guess). Before they hatch, you will find them as small, thin, red worms (larva) in the mud bottom of the lake. We commonly find them in mud on paddles and push-poles used to hold or push the boat. During the day, adults are also dined on by dragonflies, tree swallows, and other birds.

Take a nice continental breakfast or sandwich, a couple of cool drinks and a pair of binoculars with you. If you should find a radio on the boat, hide it or sink it overboard (not to be a litter bug, retrieve it before you leave). These mornings are usually very still, but if you experience a hint of breeze, go upwind to a good location and drift downwind taking in all of nature's best. Take along a light anchor (in case the breeze picks up or you wish to stop for a while) and plenty of refreshments. For the next hour or so, you will be amazed and greatly pleased that you made the trip.

Temperatures usually begin to fall and become pleasant or tolerable on the lake the last two hours before sunset. Repeat the preparation procedure in the sunrise cruise, except begin two hours before sunset and stay until the last of dim light. A pontoon boat or a good size flat-bottom boat works very well. Be sure the watercraft is rigged for regulation night travel. It is not necessary to go very far from shore, or from your dock for a pleasant experience. Just allow a good stretch of reasonably open water between the camera and the subject for reflection of the setting sun. You might even decide to stay a while and enjoy night sounds as well. In this case, you might as well go for the max; take plenty of cold liquids, a little meat with fixins', and a small charcoal grill. Keep it simple; you will create

Figure 56. A native wild hibiscus flower.

one of the simple pleasures of the great outdoors. Watch the sky, as a continuous movement of egrets and herons are heading in one direction; very likely, these are parents returning, perhaps from the river, with food for their young.

Late Summer (Mid-July—Mid-September)

Mississippi River beaches during dry seasons can be as knock-your-socks-off beautiful as a Gulf beach. Very few take advantage of this neat outdoor opportunity. The Lower Mississippi River floodplains inside the levees can seethe from summer heat after the fourth of July. Giant ragweed can grow fifteen feet high, and Johnson grass ten feet. Temperatures frequently reach and exceed ninety degrees in nearly 100 percent humidity; it's a jungle, and feels and appears as tropical as the Amazon River. Some years droughts dry up nearly every wet place in the country; the river runs low and clear. Reelfoot Lake, within sight of the river, has dried enough that my father and his sisters walked across the head of Upper Blue Basin from Walnut Log to Gray's Camp in their street shoes. Can't do that ordinarily; you'd need to walk in three or four feet of water and two feet of muck. Rough years for crops, I'm certain, but dry summers are part of nature's faultless scheme for ultimate harmony within the ecosystems, and good outdoor opportunities.

A dry fall, which is common, is one of the most dependable times to find beautiful beaches on the river. It's also the most pleasant time to be there. Besides all of that, it's one of nature's essential therapies for native rivers and their wetlands. Arid conditions might seem an uninviting time to be out during hot summers, but don't go into senescence as some plants

Figure 57. A Mississippi River view from a dike
in late summer and fall.

are inclined to do; get out and about during the cool of the day, or stay close to the shade and enjoy the experience. Investigate the dry cypress swamps, or take a short trip to the sandy bars of the Mississippi River.

Want a panoramic view of Lake County? Take a slow cruise along the Mississippi River levee. It is quiet and beautiful early and late when a low mist covers the flatland; you might not see a single vehicle along the way. We usually start a few miles into Kentucky, enjoy the scenes and soak up the history and stories of the river and floodplains to the New Madrid Bend; it all relates somehow to the creation of Reelfoot Lake.

Figure 58. A Mississippi River sand beach,
common during late summer and fall.

Thoughtful reflections also remind us that the mighty old river has a personality. Mark Twain thought so. He was a steamboat pilot on the Mississippi around the 1860s, once living not so far up the river, in Hannibal, Missouri.

Twain wrote about a number of incidents and features of the river adjacent to the wetlands of Reelfoot from Hickman, Kentucky, around "Madrid Bend" (Kentucky Bend) to Tiptonville. But he made no mention of Reelfoot Lake. An excursion on the Mississippi, however, helps us to appreciate the entire history of the floodplain and its association with people and the river.

Area crops have been "laid by" this time of the year; hardly anyone is around, and across the floodplain is a lovely lush green with the fresh fragrance of chlorophyll—and all is quiet. Deer, still wearing their burnt-orange coats of summer, begin to come out of the thickets to the edges of the fields, partly to escape the deer flies and partly to grab a bite of soybeans. An occasional bald eagle passes overhead, sometimes followed by a brown-feathered juvenile. A bobwhite quail might run ahead, darting in and out of the "weeds" (essential to their survival, assuming a bush hog has not mowed them down).

The river bank often seems to be the most pristine wild land in the county. Habitat here is usually natural and diverse. Except for the sand dunes and willow bars, the landscape is not marinated with pesticides and plows rarely go there. The remote places and the levee are about the only habitat remaining to support this vanishing gamebird. The point here is that a diverse habitat is required to produce diverse native "weeds" for food, cover, and nesting habitat for native animal life—and this habitat has become perilously rare. A diverse native weeds habitat produces diverse populations of native "bugs"; diverse insect populations feed diverse populations of native birds, and all of it together supports fish and wildlife—and humans. And that is one of the purposes of Reelfoot Lake. If we continue our passion for nature and hope to pass it along to the next generation, we should not forget these simple truths because they apply to everything we enjoy about nature, and the lives of every living thing. We are becoming wiser about the management of wetlands, wise enough to improve natural treasures like the complex ecosystem of Reelfoot Lake.

Back to the river; its channel is mostly out of sight from the levee road. You need to go about a hundred yards or so beyond a thick belt of water maple, black willow, and cottonwoods to see it. Along the way, you will find a lot of scattered drift wood and debris collected from the North Country when the river is high.

Figure 59. Bobwhite Quail, once abundant in Lake County.

While enjoying this tranquil trip, you might as well think a little about reality. Imagine a flooding Mississippi with a swift-flowing current only a few feet below the top of the levee. That is the way it was during the spring flood of 2011, and again in 2019. Was there a weak spot anywhere along this stretch then? If this was a hundred-year flood (history has not forgotten the flood of 1883 and dozens of others), what might the levee withstand in a thousand-year flood event? What would the wetlands of Reelfoot look like after the next great flood—good or a disaster? Could we withstand the consequences? It's hard to say. Everyone in the river floodplain thinks about these things. But the love of living in the river floodplain is in our genes now, and however the river rebels, we are likely to forgive it, remember its generosity, and hope it forgives us.

Continuing south along the levee, there's a little skip when the levee reaches the small community of Cates Landing (birthplace of Major General Clifton B. Cates, a decorated Marine of WWI and WWII, and later Commandant of the US Marine Corps). It's on high ground. You will notice the promise of a major river port here. A short stretch of the most modern kind of highway in Lake County is at the entrance to prove it.

Approaching the next bend in the levee is the "squeeze," or neck, of Kentucky Bend. Go a mile across this neck and you might find yourself twenty miles downriver. This is about the neck of Kentucky Bend, about where the river is tempted to jump across the narrow stretch of floodplain for a shortcut. If not for the levee, it would have already done so. It leaves one to wonder, "Why have we prevented the river from jumping across this narrow neck of land? Would we not have far less anxiety about the river if we simply allowed it to make its own corrections?" Of course it might have isolated part of the Bend without some other bridge way. It certainly would have created a huge natural lake about twenty miles long.

We get off here, a little beyond the thin neck of Kentucky Bend, where this stretch of levee ends, and head back home. The thawed bluegill from our last fishing trip should be ready to roll in meal. Pretty soon, they would be on a platter, fried crisp and brown with a tall glass of iced tea and side dishes of baked beans and cold slaw.

Great blues are always entertaining. We watch one for hours fishing from a stump off our boat dock. It is his territory and he might claim it all summer—maybe even in the winter. These huge slate-colored birds can usually be found here year round, but late summer is prime time to watch for them as they are often the most common bird you will see. Great blues defend their territory with fantastic zest and bravado. They are loners that might be the most entertaining bird on the lake; watch as they patiently stalk prey and fuss with their neighbors for territorial rights. Intruder birds will be met with a flogging and the most erasable volley of squawks and other verbal abuses imaginable. I've mentioned there are at least two major rookeries on the lake, very remote, where these birds—along with egrets, anhinga and other large wetland birds—nest and raise young. I've also mentioned that park naturalists or rangers often provide escort for small groups to these sites. Most great blues migrate south during winter, but some seem to stay around until freeze-up.

Figure 61. Our dog Wynette hiking a Grassy Island Trail and board walk.

Hiking trails and boardwalks can be the most enjoyable and rewarding opportunities at any park, and Reelfoot Lake is perfect for hundreds
of them. These facilities are not plentiful at the lake but there are a few.
Cool mornings and late afternoons to beat the late summer heat can offer a pleasant break from routines. One red flag to mention—deer flies!
Hatches in late summer can be an unpleasant experience for hikers. Deer
often retreat to open spaces just to escape deerflies. It is easy to test for the
presence of a hatch—step under the shade of any large tree! If a deer fly is
not buzzing your ears, you're probably OK.

Beautiful white birds, American Egrets are an outstanding feature
of the lake. Trees full of snow-white egrets gathering for roost is a scene
not easily forgotten, and they are here all spring and summer. Their
nests are often found in the same rookeries as herons and other egrets.
It is breath-taking to watch hundreds of these birds gather around
sundown to roost in the isolated cypress groves still living in the open
lake. One of my most vivid memories during my youth is sundown-
summers and egrets. Setting and baiting trotlines for catfish is usually
done around sundown to prevent bluegills from knocking off the bait
before night-feeding catfish come along. It is a perfect time to observe
egrets gathering to roost.

The scurry of roosting egrets can be hazardous to a trotline fisherman. It takes concentration to put tiny baits on sharp hooks, in a breeze
with a moving boat, steal a glance at the breath-taking scene of an egret
roost at the same time, and not get a hook in your hand. A grove of green
cypress weighted down with hundreds of snow-white egrets can't be

Figure 62. Egrets on stumps drowned by New Madrid earthquake.

ignored. Streaming from the sky in many directions are hundreds more, coming with expectations to find an empty space in the trees to join their cousins. Spotlighted by the soft, orange-red rays of a setting sun (or against the dark clouds of an approaching thunderstorm), the ethereal scene is indelibly printed on these memories of my youth.

Egrets are also inclined to gather and fish from exposed stump fields during low lake pool. One might see hundreds of these birds in one setting, often near their roosting sites, calmly fishing from their individual perches.

One of the greatest celebrations at Reelfoot Lake is an annual duck blind drawing held the first Saturday in August. Three hundred or so eager hunters gather to draw for sixty blind sites. It is tantamount to a state holiday for duck hunters. It might as well be, since work day or not, Reelfoot duck hunters are going to show up. Successful participants will spend the rest of summer and fall installing new blinds or pampering what is left of it from the last season. For a fifty-dollar fee, successful applicants will have possession of the blind site for one year.

Fall (Mid-September—Mid-December)

Fall is a wonderful time to be at Reelfoot Lake. Unless an unappreciative storm rolls in from the north and interrupts the master painter of fall leaves with strong winds and freezing temperatures, the scenery around the lake will be electric. Cypress trees become a gorgeous copper-bronze, too beautiful to describe. Cypress trees are definitely star performers, but not the end of the show; the brilliant yellows of hickory and water maple, the deep reds of sugar maple and sumac, all set the background of a gorgeous fall along the bayous and canals. Reflections in the lake along the shoreline are frame-filling shots. In a boat, on an auto tour, or down the trails and boardwalks, you will long remember the fall day you spent at Reelfoot Lake.

Do something a little different and take a walk through a dry cypress swamp for another otherworldly experience. For the more adventurous, walk through the hardwoods at Black Bayou Refuge to any of the cypress groves in the vicinity of Crane Town, Big and Little Ronaldson (include water, a snack, a map and compass, and a cell phone if you aren't a seasoned hiker). Otherwise, get a park ranger or Gumbooter to take you into the area. Walking among cypress knees as tall as a six-foot man, and stretching out for a short break in a thick carpet of cypress leaves is probably unlike anything you've done in the Southeast. Trees in this

Figure 63. Joe Guinn stands among tall cypress knees.

swamp are the healthiest you'll find, many taller than a one hundred feet. This is how nature planned a cypress swamp: it floods during most winters and springs and dries during most summers and fall. Trees need dry ground to thrive; they cannot sprout and grow in permanent water; they die. Cypress trees standing in the open water at Reelfoot, unlike these, will eventually starve for oxygen and succumb to permanent flooding. Judging from the open water cypress on Reelfoot, some will survive more than a century. Cypress trees are just unusually resilient and have a great tenacity for life, so they hang on.

While you are at it, take a fall trip along the Chickasaw Bluff. There are different routes to get on the Bluff Road; one is from the Reelfoot National Wildlife Refuge Office on Highway 157. Going a short distance from the headquarters to Walnut Log, the Bluff Road bears right and follows the bluff almost to Hickman, Kentucky. Or, if Fishgap Hill is passable, take a right off of 157 about a mile past Walnut Log; the road "T"s at the Bluff Road. Along the way, especially during the fall, a few openings in the trees provide an eagle's view of Reelfoot Lake. One year we happened to have a three-inch snow in late October, Halloween Day. Only a picture can describe snow on colorful fall leaves.

While you are here, look west through an open view in the trees; you can see Reelfoot Lake and the Mississippi River. Remind yourself that the toe of the bluff is the eastern limit of the Mississippi River floodplain as well as the boundaries of floodwater prior to the mainline levee. So, as I mentioned earlier, the flooding river without levees once had a tremendous expanse of floodplain from east to west (twenty miles and more wide, in some cases) to dampen its energy. Consider also that the depth of flooding would be significantly less, and so would the detriment to human developments. The lesson is that the trade-offs of benefits and risks must be carefully assessed before overpowering the will of a river. This view is one of the best to perceive the value of a wide but levee-restricted floodplain.

Short state-wide waterfowl seasons (split seasons) are scheduled in mid-September and November. Most crops have been harvested, and nearly all sounds of harvest time are silenced. From now until December, waterfowl hunters will be busy refurbishing duck blinds in fields and out on the lake. Between the short seasons, all is fairly quiet on the lake until the long state-wide waterfowl seasons that begin the first Saturday in December.

If you see small, lovely acrobatic birds, dark above and white below, you see tree swallows. Large flocks of these birds begin to spend a lot more time in even larger groups than you saw during the summer. They begin to gather in October for fall migration. The air along the lake shore can

Figure 64. Fall on Fishgap Hill,
Chickasaw Bluff.

Figure 65. Snow geese, Black Bayou Refuge.

be thick with them gleaning the air of insects for two weeks or longer to increase their fat reserves. Power lines are weighted with them, and a few might still be seen gathering in the low trees on the lake. By November, I see very few tree swallows.

White pelicans might appear as white boats cruising at the far end of an open basin. The size of a white pelican can be that deceptive. Fall movements of these large birds have once again been announced by frequent sightings of large flocks rafting on the lake. Although a little unpredictable, Keystone Pocket at the south end of the lake is usually an excellent viewing station since State Park has a nice pier here, and the

Figure 67a. White pelicans, Brewer's Bar.

pelicans frequently congregate offshore. Pelicans, not too long ago, were rare birds at the lake. Recently, State Park has made a special event upon their arrival— the Pelican Fest in October. Like the spring migration, we see hundreds in a flock, and sometimes thousands. Park rangers often announce special tours to enjoy the arrival of these birds.

Hiking and camping are never better than they are from late September through mid-December. Usually dry weather, warm days, and cool nights are the norm. Fresh air and whiffs of kindled campfires, coffee brewing, and the tinkle of dishes fill the atmosphere at campsites. And the yip of a happy puppy accompanied by low conversation and laughter are nostalgic to anyone who has camped before. But there's a certain bonus when leaves are in fall colors. Campgrounds are likely to have plenty of spaces this time of the year at the lake, which means someone is losing out on a fine camping trip. Campsites at the south end of the lake fill first, presumably because more dining and grocery facilities are handy, or they simply like to be with the crowd.

Seasoned campers like my friends the Robinsons and Sherwoods have rarely missed a spring and fall camp for many years. The campground is right at the lakeshore and surrounded by nature (two miles of good trails). There's elbow room and a boat ramp at the end of the park. Setting a few yo-yos and limb lines on arrival day will almost assure dinner menus with catfish as the entree. A yo-yo is a spring-loaded metal device that looks and acts something like an honest-to-goodness yo-yo. It has a trip spring that releases a wound-up wheel . . . ZIP! And line and hook are snapped up with your prize; another fish is in the cooler. A plain limb line is simply a stout length of line about three- or four-feet long with a No. 2 hook. Yo-yos can be attached to any branch, strong and stout, or limber; limb lines are secured to a limber overhanging branch with the baited hook two feet or so in the water. Big game and small game hunts are annually scheduled on WMAs and refuges. These are usually draw-hunts that assure the highest quality and most enjoyable in Tennessee and Kentucky.

Winter (Mid-December—Mid-March)

My calendar for winter was the ambient elixir of Bruce McQueen's kitchen— coot and gravy, and black coffee hovered in a low mist (like a godly offering of good-eats) over the quiet waters of Bayou du Chein. The cook was Mr. Bruce's wife, Annie Lee. With a full breakfast already under my belt, boat loaded, and on the way to the lake in the dark, it would be ironic

Figure 68. Bruce McQueen's Old Home Place on Bayou du Chein.

that I'd be invited for another breakfast. Yes, but I could expect that if Bruce saw me passing. That would have been the second morning of duck season (the first day of winter in my unofficial mind) because Bruce's coots were traditionally and legally harvested on the first day of the hunt. Bruce was the brother of nearly a half-dozen others at the lake, and uncle of twice that number. All of the brothers were philosophers, some commercial fishermen, guides, store owners, and merchants of some sort. Bruce McQueen had those traits as well, but he excelled as a weatherman, a US presidential and Mississippi River historian, and he had an Oppenheimer mind about the making and use of the Atomic Bomb. He and Annie Lee lived across the bayou at Walnut Log. Motorboats had not yet arrived at Walnut Log, so the loudest noise was generally the soft "click-clack" of oars coming or going, or the bump on a wooden boat while someone prepared to launch along the bayou. During warm seasons, Bruce would be on the porch sitting in his rocking chair a little after sunup puffing on a chin-stem briar pipe. If possible, he would have you sitting in the empty chairs made available for guests and longer conversations. If it was mealtime, you'd certainly be invited to sit at his table. But duck hunters were an anxious lot, and he had little luck delaying them for very long on the way to a duck blind.

During winter Bruce would be up before any duck hunter with coffee on the stove and a rocking chair by the window with the porch light on.

From here, he could easily see boat traffic on the bayou, or anything else happening in that view. The first thing he usually noticed during duck season was the flicker of your headlight coming his way. He had plenty of time to throw on a coat and step out on the porch, freezing or not. Just before your boat entered the reach of the porch light, he'd inquire about your person, "Hey! Who am I talking to out there in the dark?" Surely he knew you because he met no stranger.

It was traditional the first day of the waterfowl season to hustle up ten fat coots and have them lying on Bruce's front porch at the end of opening day. Ten coots was not a tall order; an amateur could bag a few coots on opening day of the season. I enjoyed doing that for a neighbor. It was Bruce's habit to confirm his order, so he would be especially watchful early morning on opening day to affirm the tradition. Satisfied, after hailing me on the way to the lake, he would wish me luck, and with a wave of hand, send me on my way.

Etched in my memory to this day is his famous menu the morning after: coot and milk gravy with coot drippings, and eggs sunny side up. Side dishes were "cat-head" biscuits, fried potato patties, sorghum molasses, and strong, black Maxwell House coffee.

The likes of Bruce and Annie Lee McQueen's generosity was not so unusual at the lake, but you would not find any to surpass their humanity. Mom and Dad were like that, about the same folks you'd find congregating for a little conversation at the small stores and docks all around the lake: Miller's Camp, Hamilton's Camp, Nation's Camp, and the shoreline of Samburg; Blue Bank, Cypress Point, Gooch's, Gray's Camp,

Figure 69. Mallards on ice waiting for thaw.

Figure 70. Winter freeze-up at Gray's Camp.

and a few others. Anyone who hunted ducks at Reelfoot long remembers the excitement of those frigid mornings. But it was the waft of fried coot, coot-gravy, and black coffee on the bayou from Bruce's kitchen that I remember most vividly—the beginning of winter. Today, duck-hunting season opens with a different twist: it begins at about 3 a.m. From that time until slap dark, the prospect of a peaceful day at Reelfoot is out of the question. And it continues until the end of the season. I've mentioned this before, but here's a better version.

The quiet morning is first interrupted by the slamming of screened doors, clanking on aluminum boats, and the yelling at energetic retrievers. Next are the tires on gravel roads and the roar of pick-up trucks and bouncing boat trailers in a rush to the boat ramps. Excitement is in the murmur of voices and hustle and bustle is everywhere. The Doppler effect of whining motors quickly fades into moans as these watercrafts stretch the distance between us; a tentative quiet doesn't come until near daylight. Then you know for sure, it is duck season. If Bruce McQueen's coot n' gravy breakfast is the wakeup call of winter, shotgun volleys at first light is a signal that opening day of waterfowl season has arrived. The roaring crescendo actually begins minutes before official daylight, and lasts for half an hour. Then it settles down. But the marshes will have no quiet until the 3 p.m. daily closure is in effect. Echoes off the cypress timber around the lake sometimes seem louder than the guns.

Then, for a brief time, the hustle and bustle of hunters and watercrafts starts again—it's the 3 p.m. closure; clean out the blind, the empty shell hulls, coke cans, half-eaten sandwiches, bones of sorts, empty snuff cans and shell boxes, and so on. Then, it's load up the boat and buck the white-caps to get back to the launch, repack and load for home, clean ducks, cheer up the wife (if one is around and not with you), and get ready for the next day. Few can do this every day of the sixty-day season. So you will see a small army of camo- and cold-weather-dressed hunters at restaurants, stores, lodges, campgrounds, and so on. It's their time of the year. Water-fowl hunts at Reelfoot Lake have been going for nearly two hundred years.

The one thing I know that can shut down Reelfoot Lake is a freeze-up. Well, not quite. Once again, all is not lost. It's true that normal waterfowl hunting practically ceases, particularly for ducks, as there is little or no open water, and ice becomes too thick for all but the most ambitious hunt-ers to break trails for boats. There have been unusual Artic winters (like 1977), however, when everything north is frozen and water and food are practically unavailable. During these winters, it seemed every duck north would temporarily reside at the lake to rest before going farther south. When this happens, industrious guides find ways to break trails in the ice from the dock or landing to their duck blinds. They might stay up the entire night to assure the trail stays open enough to access their blind before the sun rises.

Migrant Canada geese rarely come to the lake anymore. It is the unex-pected arrival of snow geese (blue geese are a color phase of snow geese) and white-fronted geese that seem to have replaced them; there can be hundreds of thousands, nay, even a million in the fields by January. These geese are unpopular with farmers, and the harvest is much too low to suit waterfowl managers because their tundra nesting grounds are being overgrazed by their high numbers. Liberal-season hunts are becoming more popular as hunters learn how to set up and decoy these shy colonial birds. "Unpredictable and skittish" is a phrase often heard to describe these geese. They skip around in flocks, a meal here, and a meal there, as if the ploy is a conscious effort to confuse hunters and create anxiety among wheat farmers whose newly sprouted wheat fields are most con-venient grazing by these hungry geese.

There are times when taking a risk is the right thing to do; rescuing a helpless deer on a frozen lake is one. A few years back, Kathy and I were notified by our neighbor of a struggling whitetail doe on risky ice out in the middle of Upper Blue—no matter how hard it tried, it could not gain footing. Finally, as we watched to assess the options, the doe lay sprawled;

it had collapsed with its legs sprawled flat on the ice. Our neighbor, Sherrill Barker, agreed with us: its fate was predictable—rescue the poor deer or it would die from exhaustion. But temperatures had already begun to warm and the ice was risky-thin. It didn't matter; Providence had compelled us to help that deer. So Sherrill and I pulled out an aluminum jon boat as a life preserver, and began scooting it across the ice, me in front, my neighbor at the rear. Ice cracked and moaned as if to anguish at our decision.

She lay exhausted, unwilling to move at all. We managed somehow to lift her on the bow of the boat, legs dangling. But she preferred to retry and gain her footing. It was a struggle. The ice continued to grumble with no consideration for our Good Samaritan efforts. But we ignored the warnings and kept pushing and encouraging the little doe toward the shore, one yard at a time. Distance is relative when you are out on the ice; what seemed like a mile was only three hundred yards. But we finally made it to the exhausted doe.

The story would have been much longer had we crashed through the ice. But we made fair progress. Finally we were there at the blessed shore with our confused victim. Wobbling the last yard or two to dry ground, the little doe stopped dead still when she felt solid ground beneath her feet, unconvinced she would not collapse again on slick ice. But she got her bearing and made a few steps. She stopped and turned to watch us for a half a minute. As if to say, "Thanks," she flicked her tail, turned toward the woods, and slowly drifted into the bush. Mission accomplished; we'd sleep better tonight.

Time has erased most of the old traditions common at the lake during freeze-up. Running a trap lines, chopping holes in the ice to raise nets, or tracking a bobcat or mink just to see where they went was a thing of the past. Still, a fresh snow reminds me of snow cream; a gallon or two of snow cream along with fresh sorghum-coated popcorn balls was as memorable as Christmas Day. We'd pass a few snowed-in days with checkers, puzzles, and books. By the third day, nothing could keep us inside but a serious ailment. There were chores to do: firewood to round up, pecans and hickory nuts to crack and bag, neighbors to help, holes in the ice had to be cut to raise set nets, and so forth. That was then. Today, we go south to Lake Okeechobee, Florida. Soaking up sun doesn't require as much energy but the years we tolerated winter freeze-ups were the best.

A cruise around the lake is an option any day. Enjoying the picturesque scene of six-foot snow icicles dripping from giant cypress trees along frozen shores makes for a pleasant day in a winter wonderland. The

Figure 71. Sleepy bobcat. Rarely stays put for a photo op.

State Park museum is open, and one can take hours here to ponder the nature, history, and science of Reelfoot Lake. Have a thing for great fried food entrees? Traditional food at the lake is as widely known as its bald eagle tours; a view of the lake while dining at one of the lake's restaurants is a very popular reason to be at scenic Reelfoot Lake.

It might be tempting to get out on the ice if it is a year of freeze-ups. Be cautious about that—best not to risk it; been there; done that. It's high adventure but requires a little experience and a lot of caution. Some ignore caution. I'm reminded of that at times during low water when I go through Horse Island Canal and scrape the top of an old automobile. How in the world did an automobile end up here when the nearest highway is miles away? Fun on the ice has its pitfalls. An interesting book could be written about winter freeze-ups on Reelfoot; none would recommend an auto tour on ice.

Winter eagle watching is still the most popular winter activity at Reelfoot Lake, and has been for several years. This is the time when migrant bald eagles join the local resident population. Eagles are usually easy to find by taking a watchful auto tour around the lake. The State Park offers special bus tours, an enjoyable trip for the entire family. You are likely to see large flocks of waterfowl; you might even get a glimpse of a mink, deer, or a bobcat!

Chapter 13

Special Topics

Some topics and subjects are mentioned several times, but only briefly. That is usually because they are mentioned in about any season of the year. But snippets of information on various animals and activities often fall short of a complete story. I've considered that and have selected a few subjects for special consideration in this chapter. Some are animals were not mentioned in the narratives of my early years because they were not here. Bald eagles and pelicans, for example, were not here when I was a kid. Neither were beavers, otters, wild turkeys, whitetail deer, coyotes, or armadillos, although most are common today. I include more about hunting and fishing, as well, since both are frequently associated with camping, photography, and other activities.

Who would think to ask about reptiles? Well, just about everybody during warm seasons when they are near wetlands. That anxiety often begins the moment they leave the main roads. Although this is not a Peterson's Field Guide on reptiles, a good guide would be derelict not to address the subject, especially as it pertains to reptiles with dubious reputations. I'll see to that, as well.

Why they are called "rough fish" I haven't a clue. All but catfish and ells have scales, so scales aren't the reason. Nevertheless, it sets apart common sport fish, like crappie, largemouth bass, and sunfish, from commercial or rough fish, like catfish, buffalo, carp, bowfins, yellow bass (which can be both), and others that have little commercial value. To the unfamiliar, rough or commercial fish might imply a source of sardines for cat food. But commercial fish have played a large role in the history of Reelfoot; carp and buffalo fish were once practically a staple food. Bald eagles followed commercial fishermen to pick up the leftovers. Like house puppies that follow a bag of puppy chow, bald eagles impatiently soar above commercial fishermen waiting to pick up stunned culls pitched aside. Rough fish on a Gumbooter's menu hasn't changed a bit for some of us. I still enjoy fried buffalo ribs as good as any fish, and I enjoy fried ribs from the exotic German carp caught during winter. Too bad you missed Mom's canned German carp and buffalo fish; breakfast paddies are just as tasty as canned salmon or tuna.

Exotics are an entirely odd subject to discuss, but here I've open the subject again. There's little doubt that most exotic fish compete with native fish for food and are destructive to their reproduction. Two more recent Asian fish of concern have reached Reelfoot Lake: silver carp and big-headed carp. Both have the same genus name (*Hypopthalmichthys*) and appear similar, but silver carp are much lighter colored. Silver carp are the ones that have the dreadful habit of schools jumping out of the water, especially in response to the sound or vibration of an outboard motor. They can fly through the air like silver torpedoes. They are generally found in schools. It's best to avoid these paranoid schools of fish when you recognize their presence.

Exotic plants and animals found at Reelfoot Lake, like the exotic plant water willow, and insects like the Japanese beetle, usually enter our ecosystems accidentally, as harmless novelties, or even as agents of biological control (e.g., Asian carp). While these exotics could probably be controlled or eradicated when first recognized, we have been too slow to respond; before we knew it, their populations grew into a crisis. Introduced during the 1970s to clean catfish ponds, these Asian fish escaped and were in the Mississippi River by the 1980s, and rapidly expanded their range to other streams. Silver carp and big-headed carp are highly suspect (my view) as competitors for native fish foods, and detriments to native fish by feeding on their spawn. Both species are commonly thought to be only plankton feeders, but a neighbor and I caught a thirty-pound silver carp on a minnow while fishing for crappie. Since the minnow was down its gullet, I believe it was no accident; the minnow was deliberately taken for food.

Nevertheless, some biological communities seem to have adapted to a minority of these exotics. The German carp at Reelfoot Lake, for example, seems to be almost tolerated as a source of food since its introduction in the mid-1800s. The fish is very tolerant of stagnant water and seems to fill a niche often intolerable to many native fish. Their feeding habitats have an obvious downside—German carp root in the substrate of lakes and ponds for worms, insects, and certain vegetation, which stirs up mud and causes the water to be turbid and unfavorable to some freshwater plants and animals. And I see no reason why they would avoid the succulent spawn of bluegill beds and other fish spawn for food, although I do not know that research has confirmed this point.

As with most exotics, they quickly expand their range once entering contiguous waterways, and we have been slow to acknowledge the impact and need to eradicate these fish. Nevertheless, stronger regulations have been enacted to minimize the problem. TWRA has recently hired a biologist for this very purpose. But any solution will require a coordinated and a determined effort by all states and the federal government to combat the problem. They should waste no time in pursuit of this objective.

I have mentioned several times that white pelicans are recent migrants. Reelfoot is actually out of their traditional range, but the water-loving visitors have been coming to the lake for several years. This arrival tells us that their former range has changed somehow, and that they have adapted. In recent years, white pelicans have been major attractions at Reelfoot nearly every season except late summer, even though they are relative newcomers. Their population has increased ten-fold since the mid-1960s, which might be the main reason they have become common in the western part of the state. I do not know if anyone is keeping a census, but it is not uncommon to see three or four thousand in Upper Blue Basin and, perhaps, two or three times that in Lower Blue Basin.

The traditional migration route of this great bird is out of Canadian provinces, then south through the central states, and on to their wintering grounds in Mexico and along the Gulf Coast. But some are mavericks that might show up anywhere. One necessary element to their survival is natural wetlands, mainly shallow lakes and sloughs where they can herd and catch rough fish like shad. Unfortunately, they sometimes get into trouble raiding commercial catfish ponds.

We were always pleasantly surprised to see a single bird in years past, but eighteen years ago, about three thousand white pelicans came to visit

Figure 72. Fall pelicans raft in Open Lake.

us at Upper Blue Basin. They cruised only a few yards offshore, feeding back and forth mainly on schools of shad for several days. Pelicans are known to feed on about any fish, mainly schooling fish not too large to swallow, even Asian carp. The problem is that these exotic fish grow large very fast and become too large for pelicans, or predator fish, as food. We have seen large flocks in Upper Blue many times since, mostly during spring and fall migrations. But a raft of pelicans might magically appear any month of the year. In Lower Blue Basin, an annual festival event is held to watch the pelicans. Brown pelicans are rare; I've seen and photographed only a single brown pelican, found most commonly on our southern coasts.

Largemouth bass fishing was something of a gentleman's sport when I was a kid. Conventional casting was once a standard method: just chuck a favorite bass plug into an imaginary "X" in the water with a Zipco, Shakespeare, or Garcia rig and you were in business. We usually caught a good stringer of fish. Although big fish were often released after a photo session, enough were kept to have a fish fry the same evening. Sporting fisherman with more class than mine wore a slouch hat with fishing hooks and a guinea feather, khaki bush paints, and fishing vest. But the sport is even more fashionable today; the image, equipment, and techniques of former years is obsolete, every detail wiped away.

Today, the sport of bass fishing seems to be half-fishing, half-boating; a competitive sport, like race car drivers. In fact, many bass fisherman costumes are like race car drivers' bright and colorful jumpsuits. The front, back, and sleeves of the jumpsuit are crowded with product patches. Fast and classy, their boats are powered with outboard or inboard engines that compete with the power and speed of automobiles. Loaded with sonar, the boats may have GPS, cook stoves, refrigerators, Google Earth, side-view depth finders, and other things I am not yet familiar with. Top-of-the-line rigs also have high pedestal customized seats, which allow a bass fisherman to easily be identified for some distances; even their tricky style of casting techniques is an unmistakable clue. Modern bass fishermen are also acclaimed conservationists; they usually keep their catch only if it is to weigh in the poundage caught for competitive fishing tournaments. Possibly, they have never eaten a fried filet of bass.

Whatever else the sport is in today's world, bass fishing has always been a great sport in this part of the country. Many Gumbooters still rely on old methods. Standard lures for us were store-bought river runts, Hawaiian wigglers, or hula popper baits. We also made our own baits. One was a go-devil (buzz bait today), made from the tin commonly used to armor our boats, and a single tribble hook on a wire. We made skirts from dog hair, squirrel tails, or locks from a generous sister of girlfriend. But Herter's catalog had all kinds of bait-tying supplies, and we often used them. It could be cast or tethered on a cane with an eighteen-inch length of fishing line and swept over stumps, under trees, or through the open holes of "lily" fields. The shock factor from the strike of a five-pound bass is quite a thrill.

Another bait Gumbooter like for bass is a "red-bobber"—a red rag draped and secured over a large tribble hook and rigged the same as a go-devil. It had advantages; you could simply bob it up and down in thick cover, such as holes in "lily pads," dense "moss," and other tight cover where conventional methods wouldn't work. Modern bass fishermen use a lot of "soft baits," but they also use baits their grandfathers used.

Crappie fishing was not a big sport at Reelfoot Lake when I was a kid. Surprisingly to fishermen of today, we paid little attention to the sport fishing side of this regal fish. Of course, crappies are highly esteemed for table fare, and we knew that only too well. But instead of fishing for them, we took them with commercial fishing gear, usually nets (called set, barrel, hoop, or fyke nets), or trammel nets. Yes, crappies were sold as commercial fish until June 2001. Small crappie, less than about ten inches, were culled from the catch for the market because restaurants sought large "platter-size" fish. The culled small crappie turned out to be perfect-size table fish for Gumbooters—in fact, we preferred the flavor of the

Figure 73. Spider-rigging is a favorite crappie fishing method at Reelfoot Lake.

crispy-cooked smaller fish. Closing the commercial harvest for crappie was a two-edged sword for us—prohibition made the difference whether commercial fisherman could stay in business or not.

Today very little commercial gear remains around the lake. A few old-timers, however, cannot resist a few days of fishing commercially, even if they give away most of their catch. Losing these skills is not a good thing, because those who still have them could well be needed to help eradicate exotic invaders—Asian carp, walking catfish, and other exotic fish.

Bluegill fishing was top billing when I was a kid; nothing was more sporting than to hear the "zing" of line slicing the water from a feisty bluegill. We used a porcupine quill for a bobber (when we had them, or bottle cork when we didn't), a small hook, and a cockroach or "horseweed worm" for bait. You stopped gagging from the smell of roach pellets the very first time you caught a stringer of these fine fish. Horseweed (giant ragweed) worms are the larvae of the *Papaepema nebris* moth. Giant ragweed is common along roads, fields, and other sunny places. Those with larva in the stalk are easy to tell because of the swollen section on the stalk where the larva lives. Cut above and below the swell, and you have sticks of fresh bait whenever you need it. Today, the bait is a cricket, or a small jig with a waxworm. I knew of no jigs used at Reelfoot Lake when I was a kid, and giant ragweed worms were used instead of wax worms.

Late afternoon fly fishing is a lost art at the lake, but we considered it to be as enjoyable as the sport gets. A small floating popper on the end of line with a tiny black gnat tied six inches in front was the ticket. Slip out to the stump fields about the time the sun sets on top of the trees on the far shore. The swirls and smacking sounds of "gills" snatching "woolies" is the first signs that you have the right day to fly-fish. As I have mentioned, woolies in the adult stage are often mistaken for mosquitoes but they are harmless. Entomologists call them *chironomids*, tiny red nematodes before they pupate and hatch; both stages make excellent fish food but the small larvae seem too small to put on a hook.

I had only a seven-foot switch cane (a tough, limber, native river cane) for a fly rod, but it worked rather well. A soft delivery of the flies near a hollow stump was a guarantee for at least one good fish—and it was not uncommon to find a fish on each hook. When woolies and mayflies hatch, they swarm the low-hanging cypress limbs out in the lake; 'gills are likely to be below waiting for one to fall in the water. Quiet tight-line fishing with eighteen inches to two feet of line with a small hook and light-weight sinker is a good technique. Crickets, wax worms, or earthworms are good baits. Otherwise, tight-line fishing (no bobber) with a small jig and wax-worm will do fine.

When most bluegill fishermen are talking about going fishing, they mean during May and June, the main months these fish go on beds to spawn. It is not uncommon to catch a mess of bluegills from a single bed. Light line with a small hook and a porcupine quill as a float were a perfect combination for fishing bluegill beds. Before motor boats became popular, a Reelfoot stump-jumper and a sculling paddle were the equipment of choice for this kind of fishing. It was very quiet and the bed was hardly disturbed from stirring the water. Since there was no limit, coolers were often filled to the brim with these table fish.

We did not filet pan fish back in those days; we scaled them. A nice coat of meal with a little salt and pepper prepares them well for a hot skillet and oil. This was an easy and common way to prepare a fish dinner when working out on the lake. There were traditional landings or old campsites for this purpose. The ground here was nearly free of weeds and under-growth from wear, and snakes didn't care much for these sites. On Upper Blue, we went to the "Game Warden Shack," located on the east side of Long Point. Game wardens often spent the night here to do their business. To get there, you will find a boat canal at the head of Upper Blue that runs a short distance west along the Long Point National Refuge boundary; pass the entrance on the right of the First Arm Canal; Long Point woods is just ahead. Today, the landing has grown up, but fifteen yards into the woods stands the relic of the old shanty that harbored many memories for game wardens and kids at Walnut Log.

Catfishing with limb hooks, floats, and trot lines has become the most common method to catch catfish by a casual fisherman (commercial fishermen used trotlines that went for a mile and had up to six hundred hooks). Sport fishermen are currently allowed fifty hooks—check the latest regulations. Using any good twenty-pound test line (usually nylon twine) with number 2 hooks and an optional sinker, lines can be hung from limbs to depths of two or three feet with good success. Some use a mechanical device called "Yo-Yo" to up their catching success, but one can get by without them. The trick is to select "springy" limbs; if stout limbs are used, the fish is likely to pull out the hook. With Yo-Yos, however, any limb will do since the device is spring-loaded.

Catfish hooks are baited with about any kind of meat imaginable, but some are classic: crushed snails, crawfish, minnows, cut-up shad, earthworms, cut-up chicken breast covered in cherry Kool-Aid powder, dog food, some soaps, and a few others—some of which I will not mention. Some commercial fishermen, however, still use slat baskets baited with powerful-smelling cheese to catch these fish for the markets. Wild catfish filets from Reelfoot Lake, in my opinion, are the best on the market. Catfish are an under-harvested resource at Reelfoot.

Common reptiles (snakes, lizards, turtles) and amphibians (frogs, toads, salamanders) at Reelfoot Lake are abundant and a special category of wildlife. Rarely does anyone but a herpetologist go looking for them, but they are likely to be noticed anywhere in and around the lake.

There are about a dozen different kinds of snakes here; only one demands extra caution—the cottonmouth. In my experience, it does little good to describe this snake because most will end the conversation with a sour face and a wave of the hand with: "They're all deadly to me!" Some can hardly look at a reptile book without cringing. However, I will acquiesce and mention the picture here of the cottonmouth. Who could forget a face like that? The head is wedge-shaped, eyes a menacing "cat-eye," and a white stripe on its upper jaw runs from eye to throat. In the water, the entire body floats like it was made of cork (other snakes have half-submerged bodies when swimming).

Knowing your environment and the creatures living there goes a long way toward increasing your enjoyment of being outdoors. The cottonmouth is to be recognized and avoided. I know of no other venomous snake here in the floodplains, although timber rattlers were once common here. I know of only one person being bitten by a cottonmouth at Reelfoot, although the species is common. That person picked up the small snake in

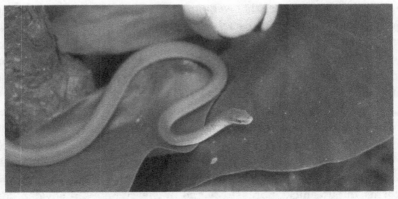

Figure 76. Smooth back green snake on Spatterdock Pad.

a turtle dip net and reached before he looked. He recovered very well—not to minimize the danger of being bitten. Just use woodsman's good sense.

The hognose snake and garter snake are likely to appear more threatening than any poisonous snake, but it's all bluff. The hognose snake (known locally as the spreading adder) will roll over on its back and play dead when overly disturbed, much like the "frightful" mud snake. The garter snake is a skinny little multicolored, striped snake that will attack with its mouth wide open. If you know reptiles and how to handle them, you will know that these reptiles calm down quickly and can be handled without more aggression. The curious would do well to procure a field guide to reptiles and amphibians. Still, as with mushrooms, avoid all reptiles unknown to you, and use caution when around them.

About five other snakes at the lake are common water snakes, plus the colorful red-bellied mud snake (Figure 35). It is locally known as the stinging snake with a mythical reputation for producing a deadly bite; not so, the sharp spine on its tail is used effectively to position its food. Harmless, the mud snake will simply roll over on its back to expose its red belly when threatened. It is an important indicator of healthy wetlands because its main foods are in the family of *amphiuma* (salamanders), amphibians that require good water quality.

Water snakes, except the beautiful mud snake, are generally dull-colored, with round eye pupils and a yellowish underside; all but the mud snake will readily bite until they do not feel threatened. I have seen even grammar school children handle common water snakes and not panic at their bite, which is hardly more than a pin prick. Still, those with a morbid fear of snakes need help to become happy campers. Land snakes, like rat snakes, hognose snakes, king snakes, green snakes, corn snakes, and garter snakes, are becoming quite rare to find, since land around the lake is intensely farmed and nearly devoid of wildlife habitat.

There are all sorts of frogs to be found throughout Reelfoot Lake: bullfrogs, southern leopard frogs, tiny cricket (chorus) frogs and spring peepers, gray tree frogs, and green tree frogs. A fresh new dimension of outdoor enjoyment is to learn the sounds of frogs; their songs are all around you during spring and early summer. All are ubiquitous within the vicinity of the lake, and all sing their breeding songs at some point from March through spring and early summer. Leopard frogs are four times larger than cricket frogs and can be found hiding in the low vegetation of lawns and gardens during the summer. Then they retreat to the lake the rest of the year. Leopard frogs might scare the daylights out of you when suddenly, unaware of anything at your feet, they leap out of nowhere. Its calls are much like the sound of a hand rubbing the surface of a balloon with arced strokes.

Cricket frogs and spring peepers liven up the nakedness of late winter and early spring with their beautiful bell-like songs. They are tiny and sometimes very colorful. The gray tree frog is nearly twice the size of the former, still only about one-and-a-half inches long. The most common place to find them is on lakeside porches and woody vegetation. It has a robust trill you can't miss on a late afternoon.

The green tree frog, about the half the size of a man's thumb and a perhaps a chorus frog, may be the most conspicuous and entertaining frog around the lake. It has a bright green body with a white streak along its lower jaw. During late summer, they are most abundant at night, climbing around night lights in search of insects. It is entertaining to put a bowl of water below the light and watch as they come for a drink, like an ordinary house pet, and return to the light for more bugs.

The American and woodhouse toads are familiar species in this area—and they look very similar. Watch them hunt bugs around porch lights and campsites; they hop around campsites, nonchalant as if they are the sole proprietors. Mostly, they are harmless to people, as we do not eat or put them in our mouths. But when these toads are handled roughly, the two external parotid glands behind the their ears excrete a milky, poisonous ooze as a defense mechanism. Pets are vulnerable, but most will avoid them, especially after the first encounter.

Bald eagles and rarely an osprey could be found here when I was a kid. Different families of this accipiter group have an odd mix of habitats in the flyway. During the 1970s, to refresh our memories, bald eagles were endangered, headed toward extinction from DDT contamination. Fortunately, since the chemical has been removed from the market, they have made a slow but remarkable recovery. Today, bald eagles might be seen anywhere, but are more numerous in Florida, Alaska, and in some lakes like Reelfoot. Midwinter bald eagle counts at Reelfoot might exceed three hundred, and more than a dozen pairs annually nest and raise young here. Naturalists at State Park's Ellington Center and Donaldson Museum often give eagle tours with an intriguing history of eagles at the lake, including rehabilitation, sometimes the use of real-time cameras, as well as interesting sightings.

Many birds, including osprey, also suffered the 1970s tragedy. Ospreys are rarely seen here during the winter, but come spring—get ready; thirty or more osprey nests might be found on the lake. They are very active

Figure 78. A bald eagle family in spring cypress.

spring through summer. Large populations, including nesting osprey, are found in Florida, probably where our summer population spend their winters. They can also be found on isolated lakes along the Mississippi River.

Reelfoot Lake is a birder's paradise. It has a large concentration of resident birds, although many may relocate a little north or south, depending on the harshness of the weather, the availability of food, and other environmental conditions. Hundreds of egrets and herons raise young here. Large rookeries are found on the lake and nearby Mississippi River, and there seems to be no end to nesting songbirds, bitterns, rails, wood ducks, shorebirds, and the list goes on.

A savvy birder can stay active every month of the year at Reelfoot Lake. Birders are reminded that the use of canoes, kayaks, stump-jumpers, and other quiet operating boats are an enjoyable and efficient means of finding and getting close to birds and other wildlife. Miles of canals, sloughs, and ditches lead to the interior wilderness of the lake. There, and along the way there, you will be surrounded by the best of the South's natural wetlands.

Birders do not cull any avian within reach of their binoculars, boat, hiking ability, or all-terrain vehicles—and their numbers are growing. US Fish and Wildlife's Partners in Flight estimates there are fifty-two million bird watchers in the United States and Canada. They spend $14 billion annually on travel and equipment. But unlike waterfowl hunters, which support their sport with licenses and permits, they have no organized way to financially support nongame birds. Yet, game birds are simply another bird to them, and every bird is on their list of sightings. Some known as "backyard birders" judiciously hang bird feeders every year not too far from a window, porch, or other viewing platform. Here, they eagerly check the feeders to rejoice at the return of old acquaintances and new arrivals. Ms. Onice Strader, a neighbor on the shores of Reelfoot, has for years kept out hummingbird feeders, maybe five or six outside her window. It is unbelievable the hummers that gather at these feeders at one time—a hundred, maybe two hundred—too many to count.

But do not be surprised to find "birders" mentioned among the most avid outdoor adventurers; they are a hardy lot, just as hardy and intense about their sport as any waterfowl hunter. Some carry cameras, some binoculars, and others just a notepad, and most have a field guide to North American birds. They often deprive themselves of worldly comforts, skip

Figure 79. Ellington center boardwalk: Trails and boardwalks
at Reelfoot Lake are great for birding.

busy work schedules and vacations, and go hundreds of miles on a weekend, just to see a single bird not on their list. They are known to endure the hardships of dust, heat, wind, rain, and blinding snow to accomplish their birding objectives, not unlike duck hunters.

Not too surprisingly, birders are some of our best amateur naturalists; they are amazing outdoor observers. These outdoorsmen and women are extremely knowledgeable about their sport, and since birds require healthy habitats, the birder rarely misses anything of note in the wild. It gives them more than a good reason to be outdoors. Their main contribution is volunteer services to conduct research, count and report bird sightings, or assist through the Audubon Society, Partners in Flight, the National Wildlife Federation, and other conservation organizations.

Whether you are a birder or not, a quiet trip through the many canals, bayous, and basins in a canoe or Reelfoot stump-jumper, any season of the year, can be an unforgettable adventure. The navigation routes are replete with a diversity of habitats—from upland hardwoods, stump-studded basins and fields of giant cutgrass, button bush, and pickerel weed to "lily-pad" marshes. Logs, beaver dams, and cypress domes provide stations for naptime, lounging, feeding, and nesting for a variety of mammals and birds. Some of these are herons, egrets, eagles, prothonotary warblers, blue-gray gnatcatchers, spotted sandpipers, and many other birds. Mammals like white-tailed deer, mink, beaver, raccoons, bobcats and others could show up at any time, prowling the shores. Will we see this diversity of birds on their next migration?

Figure 80. A female Redstart on Equisetum during early spring.

The presence of an active birder might well be an index of world health since they would not be present if there were not a bird to sing its songs. Would either even pass our way if great wetlands like Reelfoot Lake should disappear? And where, one might ask, would all other native animals find their retreat? It will be a sad day for the world when birders no longer have a reason to go to the fields, woods, and wetlands.

Chapter 14

Auto Tours

Forty miles of roads circumvent the lake, with a dozen or so side roads to the lake, into the woodlands, or other good observation areas. You will notice that lakeshore developments are mainly limited to the shoreline from the Samburg area southeast shore, around the south end of the lake, to Champey Pocket, the exceptions being Walnut Log and Gray's Camp. The reason becomes obvious, once you go around the lake: it's about the only land high enough with limited flooding during high lake levels. This is fortunate for those of us who enjoy the best of nature; most of it is untouched wilderness, for the wild things we love to study, explore, observe, and appreciate—the greatest values of Reelfoot Lake.

Champey Pocket to Black Bayou Refuge

One of the most panoramic views of the lake is at the Champey Pocket (Champion Pocket) boat ramp. From here to Black Bayou Refuge are several side roads to the right; most go to the State Woods, a wide strip of WMA forest between Highway 78 and the open lake. Birds and mammals might be spotted any place along the way. Watch closely at side ditches with water, edges of field openings, and the canopies of trees. Bald eagles, deer, furbearers, bitterns, egrets, wild turkeys, and neotropical birds are but a few animals to be seen. Those with botanical interests will find the entire circuit intriguing, since a great variety of plant communities flourish around the lake.

Black Bayou Refuge has roads and wildlife viewing towers open to the public year round; access to the interior of the area, however, is closed during waterfowl season. During spring, a great variety of waterfowl, wading birds, shorebirds, egrets, and herons can be seen, and an occasional sighting of bitterns. From December to late February, you are likely to find swarms of waterfowl, especially at sundown, and bald eagles looking to have a duck picnic. Go on north over Highway 213 to Blue Heron

Road and a short quarter-mile to the refuge's Phillippy Unit, where you will see much of the same.

Gray's Camp/Air Park Campground

Stretched some two miles along the western shore of Upper Blue Basin, this small historic community still has a few buildings with a flavor of the old Gumbooter days. Today, it has two public boat ramps, three or more resort businesses with guides, lodging, and boats, and a short dozen resident houses with several more weekend camps. At the end of Highway 213 is the Airpark Campground. Along the entire shore is one of the most scenic vistas on the lake. The campground itself is scenic enough: under a closed canopy of oaks, pecans, cypress, and maples, two miles of loop trails supplement the setting, some along the wooded shoreline, some through old field succession, and some through the campground. Signage may be a bit short but the trails have been pretty well kept.

Long Point Unit of the NWR

Continue north on Highway 78 after leaving the Gray's Camp area; you'll cross the Tennessee/Kentucky state line after about three miles. Take a

Figure 81. Winter day treat on Reelfoot trails and boardwalks.

right on the first paved road a couple miles farther; the National Wildlife Refuge entrance is on the right. Like the rest of the refuge, Long Point is open to the public except during winter months when peak waterfowl populations are present. During the closure, a wildlife watch tower is still open to the public. Depending on the uncertainty of the weather and production years, you are likely to enjoy an hour or so watching thousands of waterfowl come and go at the refuge.

Walnut Log/Grassy Island NWR

Walnut Log is an historic community mentioned frequently in this book. You will notice that the west side of the Bayou du Chein has no developments—that is because, in 1986, TWRA compensated owners of ten houses there and reclaimed the lands as part of the WMA. The Walnut Log Road is open to the public year round to the end of the road on Grassy Island, unless the NWR has special needs to close their part of the road. Once again, the refuge part of the road is as close to a public wilderness road as you will find at Reelfoot Lake.

The area of Grassy Island is no longer "grassy" but mostly upland hardwoods and cypress swamps. It makes up perhaps half of the Reelfoot

Figure 82a. Barrowed Owl: A wet
and disgruntled chick.

Figure 82b. Cardinal flower.

federal refuge acreage. Nor is the refuge an island, although much of it was surrounded by water at one time. During that time, the interior was a sea of sawgrass marsh with a several small open basins—the reason for its descriptive name. You will find these open areas on some old maps. Gradually, much of the area has been built up by sediments, mainly transported by Reelfoot Creek. Chickasaw Bluff has a deep layer of fine gradient loess—windblown soil gusted in from the west—and it is highly erosive. Heavy rains wash these soils from gullies and exposed farmland where it finds a way to the lake. Grassy Island has become a major filter for sediments and other pollutants entering the lake from the eastern watersheds.

Grassy Island harbors not only every species of wildlife on the lake, but is truly as great an experience in wilderness/cypress swamps as can be found. Two boat launches, one at Walnut Log and one at the end of the Grassy Island road, provide access to the lake. Special regulations, however, apply to the Grassy Island ramp during the winter waterfowl migration. Stop at the Reelfoot NWR office on the way back to Highway 22 and enjoy their small museum.

Kirby Pocket/Samburg

Notice along the way south a few places on the west side of the road where fields have been reforested up to the highway. These wooded buffer zones, land purchased and reforested during the 1980s and 90s, are important filtration zones. They capture sediment and agriculture's chemical runoff that threaten the lake, and minimize conflicts with lake level management and private lands. Kirby Pocket (known also as Middle Landing) is managed by State Parks. All of these state lands are open to the public—birding and sightseeing are popular activities here. It is also a popular boat launching area.

Kirby Pocket to Blue Bank is a zone of magic for sunsets where you will discover some of the most sublime opportunities for sunset memories or photographs. It is also a favorite for light watercraft trips on windy days, such as canoe and kayak trips.

The only incorporated town on the shores of the lake, Samburg is also one of the oldest. Like Gray's Camp, it is built on the lakeshore with the most scenic views of the open lake, but Samburg has sunsets and Gray's Camp has sunrises. With a full complement of visitor facilities—including dining, motels, guide services, and RV parks—visitors will find a quiet and scenic shoreline, communities reminiscent of the 1940s.

Figure 84. Fishermen's success wading among giant cypress trees
and from the boardwalks.

Samburg/Blue Bank State Park Campground

Along this beautiful shoreline is also the most residential area around the
lake. Majestic cypress trees line the shore along with the homes, rentals,
and weekend houses. At the end is a full service lodge, including guides
and boats, next to which is the State Park's eight-six site, full-service campground. Located in a scenic wooded area, it is often at capacity during
peak seasons.

The Spillway/Cypress Point

Only half a mile down the road is the lake spillway complex, the only controlled outlets for the lake and the original outlet of Reelfoot. In fact, the
area is known as "the spillway." The entire area is scenic and exempli-
fies progress for public use and enjoyment. The old spillway structure
was completed in 1931. It was the first and a key facility on the lake for
the management of water levels—the Achilles heel of Reelfoot water level
management and aquatic health. Natural lakes require no management—
only protection and use. But Reelfoot lost the ties to its parent, the Mississippi River, which once nurtured and sustained the lake's naturalness

through an ingress and egress of fish and water. That relationship was severed when the levee from Tiptonville to Hickman was erected in 1931.

The separation of lake and river began to take a toll on the lake's vibrant ecosystem within a few years. While the biological vitality of the lake was still great, my dad and granddad began to complain of signs that the lake had lost much of its former vigor; fish, furbearers, and migrant waterfowl were not quite as robust as in their day. One reason is that the lake was showing signs of accelerated aging in a process called "hyper-eutrophication." Hyper-eutrophication is a major concern with artificial lakes or where natural lakes like Reelfoot have manmade spillways or dams. Biologists have known since the 1940s, and earlier, that Reelfoot was accumulating too much rich organic material (detritus—dead plants and animals). Like too much rich compost in a garden, this material caused more fertility than natural lakes need. Because the river was no longer connected to the lake, the buildup of organic material was no longer oxidized by tumbling water or flushed out of the lake.

Water levels dictated by past policies of regulating the spillway resulted in relatively stable lake levels. Like a farm pond, the lake was not controlled by a natural stream. Such a static aquatic system causes a biological decline in wetland habitat because nutrients from dead plants and animals, runoff from livestock, commercial fertilizer, or some other source are not readily oxidized from lack of sunlight and free oxygen. Soon it becomes apparent that the limited free oxygen in the water needed by aquatic life is used up by the decaying nutrients. Consequently, over time, pond-like wetlands become unsuitable for aquatic life, which is why artificial or modified natural aquatic ecosystems depend on us to take care of them. Left unattended, most ponds must be drained and renovated about every ten years to support a fishery.

A scenario similar to farm ponds (a man-made wetland) might be the best example of what happens when a natural wetland like Reelfoot Lake is cutoff from its parent—and that is always a native river. Without the care of nature, it eventually ends up with poor water quality and a general decline in populations of fish, amphibians, furbearers, birds, reptiles, and so forth. In fact, the entire food chain that feeds the lake ecosystem is compromised. Residents around the lake and its users are aware of the trend. In spite of its unparalleled aesthetics and great biological diversity, newspapers have reported complaints about a high density of emergent aquatic weeds and low sport fishing results at Reelfoot, at least since the 1950s.

So Reelfoot has lost most of its ties to the Mississippi River. Beginning in the 1980s, TWRA made an effort to compensate for the lake's lack of integration with this great river. One step was to modify the scheduled water level management of the lake. An exceptional action was to implement a major lake drawdown May 27, 1985, whose purpose was to duplicate the effects of natural droughts and wet seasons, similar to the flooding and drying of floodplains in natural lakes and rivers. Floods and droughts, to some degree, are nature's most powerful effort to reconstitute organic sediments of a floodplain or lake bottom to common soil, which utterly revitalizes the lake's ecosystem. The anticipated results are a better spawning habitat and more available free oxygen for fish and other aquatic life. Finally, an entire wetland ecosystem benefits by this management event, providing a healthier, more robust food chain and a vibrant ecosystem. You would suppose correctly that this happens every year in natural rivers and lakes.

But the 1985 effort didn't work: the outlet gates were closed and the drawdown ended after the lake was lowered 2.5 feet in mid-July. Two major reasons explain this failure: the old spillway was inadequate to lower the lake fast enough or low enough; and the public was opposed to the idea. They were not convinced that the proposal was supported by natural principles. They were skeptical from the standpoint of inconvenience, economics, and fear of disastrous results—boats could not be moored traditionally, tourist trade would drop, well water would be tainted, and diseases would result. Raccoons and other animals, some said, would get stuck in the mud, die and spread diseases to their children, and so on.

Anything we build is destined to suffer the Second Law of Thermodynamics—it begins to deteriorate the day it is created. The old spillway had begun to show its age and was leaking badly through the unsettled controversies of the 1980s and 90s. Federal court action bogged down the process by requiring an environmental impact statement. A full decade later, the guideline results had still not been fully implemented. So out of the state's proposal, citizens lost the benefit of a drawdown and changing lake levels but gained, among several benefits, two very major things: a new spillway, which could be extremely important to the future management of the lake; and four-thousand acres of new, mostly forested land, around the lake as a protective zone to reduce conflicts between high water and farmers, and other private landowners, and a vegetated filter zone to help protect the lake from excessive sediments and agriculture chemicals.

The new spillway became operational in 2018. Water level management, in the absence of influence from the Mississippi River, remains the most hopeful and powerful management tool to restore the lake's vitality. When and how those techniques will be done depends mainly upon the public's will to do it. In fact, the uprising caused by the 1984 drawdown proposal resulted in more interest, and provided more improvements for Reelfoot Lake than all accumulative efforts during the last hundred years (since it became public domain in 1914). The sooner this becomes common knowledge, the sooner the public will insist on the best and most practical natural resource management available to Reelfoot Lake.

Cypress Point is a cluster of facilities, including a boat dock, popular restaurant, and another park facility known as the Blue Bank Round House Park. It was built, with all of the nostalgia of Smoky Bear, by the Civilian Conservation Corps during the 1930s. Many family picnics have been held here since that day. It has shelter facilities and could very well be the most pleasant and scenic view on Reelfoot Lake.

Ellington Center/ Keystone Pocket Park

Next to Round House Park is the area known as the Ellington Center. Besides the National Guard and Bo's Landing facilities, the park headquarters office is here. Bo's Landing has been an excellent location to observe bald eagles and our white pelican visitors. Park facilities include the Ellington Hall Auditorium, the R.C. Donaldson Memorial Museum, a very nice boardwalk, and the nature center.

Keystone Pocket Park is the last stop for park facilities along the south shore. The park has a pavilion and boardwalk pier to accommodate day use. Along the shore is a one-way trail. The view of the lake is scenic, particularly when thousands of white pelicans or snow geese are rafting only a short distance offshore.

Chapter 15

The Outlook

The Prospective

In a cheerful book like this one, it is tempting to avoid sensitive subjects that tend to diminish the enthusiasm, but some must be addressed. They have to do with the future of Reelfoot Lake, the people living here, and the visitors who come to enjoy nature. Hopefully, these issues will be kept in mind by those who have a passion for nature and see the importance of preserving it, be acknowledged by politicians, be revisited with zeal by conservationists, and be useful to planners for generations—and those to follow in our footsteps.

The river owns the floodplain. It is the primary creator of native wetlands like the former oxbows of Reelfoot Lake, although it sometimes gets a little help from other natural sources, like springs, seeps, and the New Madrid earthquake. The entire area of the Mississippi River from the Chickasaw Bluff in the east to the high ground west of the river—twenty miles or more wide—is the river floodplain, which serves as a reservoir to efficiently accommodate floods. It has one of the richest ecosystems we know in North America because of its deep alluvial soils that produce not only domestic crops and native wetlands but also the fastest growing hardwoods on earth. The Mississippi can be a relatively clear and lazy stream with beautiful sand beaches during late summer and fall; it can also be a raging torrent—and it objects to being constrained by anything, especially dikes, dams, and manmade levees.

For these reasons, users of the floodplain should be keenly aware that the river is resilient and will eventually prevail in a contest with human intervention. Any plan for development intended for permanent use in the floodplain should acknowledge and expect this inevitability; it is simple wisdom—use of the floodplain should be compatible with the natural functions of the river. That is the only way to maximize net benefits, minimize disaster and government bailouts, and the only way to do the right thing. Mark Twain addresses precisely this subject in *Life on the Mississippi*.

> The military engineers of the [Mississippi River] Commission have taken on their shoulders the job of making the Mississippi over again—a job transcended in size by only the original job of creating it. They are building wing-dams here and there, to deflect the current; and dikes to confine its narrow boundary, and other dikes to make it stay there, . . . One who knows the Mississippi will probably aver [confirm] . . . ten thousand Commissions, with the minds of the world at their back, cannot tame that lawless stream, cannot curb it, cannot confine it, cannot say to it, Go here, or Go there, and make it obey.

Although Twain credited Captain James Eads for some apparent success with jetties to direct the channel at the mouth of the Mississippi, he continued: "Otherwise, one could pipe out and say the Commission might as well bully the comets in their course and undertake to make them behave, as they try to bully the Mississippi into right and reasonable conduct."[1] I have little doubt that Mark Twain would agree that the right, responsible, and equitable plan to resolve the management needs of the Mississippi River corridor would mean commissioning the US Army Corps of Engineers to implement it, and that it would be done quite well.

The benefits assigned to levees should be a high priority for Congress and the country for benefit/cost reevaluation. The criteria of credits and debits should not be limited to the economic benefits of a few enterprises, such as transportation and agriculture alone, but a comprehensive evaluation that includes the inherent values of natural resources and potential disasters in the interest of the entire nation, today and in the future. Presently, every foot of rise in the height of the levees is regarded by some as profits; but for the multitudes of unknowing that live in the floodplain, a rise could be interpreted as a dismal public cost and a pending disaster.

River highways for the transport of business and agriculture products are vital uses of the river floodplain, but not to the exclusion of millions of lives, towns, highways, infrastructure, and natural resources, which is reason enough for an urgently needed national reevaluation and prompt restructuring of river corridors. Indeed, these R&Rs are the only hope to avoid predictable disasters and, at the same time, best serve the future of the country. Some will say it is too late; we've gone too far; the task is too monumental.

Really?

It would seem that way when you consider the unintended disaster of the Swamp Act of 1850, which nearly destroyed the Florida Everglades and half of the country's alluvial river wetlands. TVA has managed to clear and grub two hundred miles or so of river floodplains for a hydroelectric dam; NASA is considering a colony on Mars. Both seem justified, but they mainly show a natural resource–conscious nation that the technology is here to do whatever is necessary to restore a society with a balance of industry and sustainable natural resources.

At the very least, disaster can be mitigated by education; but the government continues to concern itself with "progress" as if river natural resources were merely an expendable toy. Take, for example, the Clean Water Act of 1972. Valuable legislation for clean water and native wetland protection, the legislation has been compromised across the nation to elude public conflict with a single industrial exploitation—energy production first. Can't we do both at the same time?

Meanwhile, common sense tells us that current strategies for managing wetlands and rivers like the Mississippi have been long overdone and outdated. But the US Corps of Engineers is directed by Congress; Congress by the people. Congress should not feel satisfied that it has enacted the public's will when the public is unaware that there's a problem—Congress and agencies also have an obligation to inform.

Public awareness makes the difference. Someone has seen to that in Florida. The Corps of Engineers have begun to address the problem: they have already abandoned channelization and started to restore the original function of the Kississimmee River. It's a start; now the Florida Everglades. A beginning in Tennessee would be West Tennessee rivers; next, the Mississippi. The survival of native wetlands and the plants and animals they support will follow.

Meanwhile, we can begin the restoration, preservation, and management of what remains, especially along major rivers like the Mississippi, and wetlands like Reelfoot Lake. Among a great multitude of benefits, these wetlands are buffers against catastrophic floods, much the same as shelterbelts of trees are to wind and soil erosion in the prairies of Kansas, Oklahoma, Nebraska, and the Dakotas.

Catastrophic floods are as predictable as bad crop years—we know they will occur, we just don't know when. With help from the government, farmers are condemned to plan for it; floodplain dwellers are condemned to do the same. Early pioneers at Reelfoot Lake were on the right path of thinking before the levees; they built their houses on piers, flotation, or located facilities above the floodplain, and crops out of the flood-risk zones.

Cropland can be custom-managed, too. Instead of managing river floods around the agriculture industry, manage the industry around river floods. In the one-year floodplain of rivers, there's a 100 percent chance that it will flood within the year; a 10 percent chance in the ten-year floodplain, 5 percent in a twenty-year floodplain, and so forth. So potato farmers know the odds of a successful potato crop in a levee-free floodplains of rivers from the beginning. What the public does not know is how much it is going to cost them to bail out the poor potato farmer; what legitimate river users don't know is that ignoring the problem will eventually cost them their river. In a flood plain protected by levees, the odds for a successful potato crop are not as certain—man-made levees can fail any season and any year. We can expect a similar outcome for any incompatible use of the floodplain. So we should look ahead and be prepared to minimize the costly outcome.

A carefully planned compromise would be to avoid huge mainline levees that prevent overflow from entering the expansive floodplain. Have Congress and the US Corps of Engineers considered customized bearers around flood-risk facilities? There is a need to free up a great deal of the floodplain where the energy of the river can be dissipated. Then, we can construct towns and communities so they will be at low risk to flood damage, and natural resource wetlands like Reelfoot Lake can be restored.

Annual floodplains of rivers were never meant for domestic crops, or anything else that cannot tolerate annual flooding. Environmental writer John Flesher stated in an April 28, 2020, release that floods in the Missouri, Mississippi, and Arkansas river basins caused $20 billion in damage in 2019. One of the greatest benefits of wetlands and a forested floodplain is that these natural vegetated zones help tremendously to reduce the impact of catastrophic floods and distribute normal sediment deposition over a wide area, as alluvial rivers are naturally inclined. Here are other benefits that evolved over a millennium; an extremely beneficial biological ecosystem that supports fish and wildlife; uses up carbon dioxide and produces oxygen (carbon sequestration); provides a sustained yield of forest products, fish, and wildlife; provides extraordinary outdoor recreation; and much more. It is ironic that we complain about denuding the Amazon River Basin from trees and ignore our own.

Figure 85. Mississippi River levee: protected cropland, left side;
exposed floodplain on the right side.

The most optimistic future for Reelfoot Lake is that, somehow, the river can once again be included in its former ecological makeup, at least to a significant degree, because it remains a wetland oasis of major importance for native plant and animal communities endemic to Tennessee as well as to the entire Mississippi River ecosystem. The price of doing nothing about it could be the most costly and disastrous for local economics and outdoor recreation in Tennessee as well as in the nation.

The loss of critical natural resources could be the least of our concerns if we continue the US Corps of Engineers' "big levee" policies. A freshman civil engineer would have known the high risk of levees long before they were certified. What kind of levee would it take to contain a five-hundred-year flood? What would happen if a levee holding back such a flood failed? At the same time, what would happen if the epicenter of an earthquake—of, say, a 6.5 magnitude—shook the earth near the river during major floods? It should not have been a far stretch for them to think through this possibility, which leaves us to wonder why it never seemed to register with General Andrew Atkinson Humphreys (the first head of the US Army Corps of Engineers, 1866) and James Buchanan Eads as they justified the use of dikes, jetties, and levees to only resolve the transportation problems they faced at that moment.

Certainly, it took a bit of hubris and determination to build and satisfy the needs of a nation; the same drive misdirected managed to unnecessarily tear down a good portion of it. We applaud our nation's strong progressive spirit so necessary during the mid-1800s. But hubris never ends well: "Don't fret about rivers and wetlands, if we desire to raise corn on the moon, we shall have plantations of it." Such talk often fails to consider that mismanaged natural resources are finite rather than sustainable; without wise decisions and good stewardship, the outcome is often irreversible. The US Army Corps of Engineers has a large cadre of sharp engineers, geologists, and administrators, but they are civil servants, and civil servants can be pawns in the hands of overly exuberant entrepreneurs and self-serving politicians, as in practically every county along the Mississippi, from St. Paul to New Orleans.

Citizens aware of the issues have the same power to change it. But it is easy to forget. Here in the Reelfoot region, the 1908 levee failed; it failed again in 1912; the one in 1913 failed; and the one in 1927 and 1937 failed. Missouri River levees failed in 2019, and local ones were too close to failing in 2011 and 2019. Insofar as I know, these records would have been no surprise to my forefathers. Until the early 1940s, old buildings like the Walnut Log Hotel, the Tennessee Academy of Science Biological Building, the Union City Outing Club, the Blue Wing Club, the Black Jack Club, and others were wisely built on eight-foot piers, and these survived the floods. But the grandiose idea of building levees replaced good sense, especially when hundreds of miles of them were built on both sides of the river. The reason the early levees failed, gurus of floods concluded, was that these structures were not high enough or strong enough to do their jobs. So, government ordered the US Corps to build the weak levees a little higher and stronger.

This solution apparently appeased the nervous floodplain dwellers without further talk about long-range flood predictions. Confidence in levees was enough that the new building trend was to erect houses low to the ground, and often on concrete pads flat on the ground. I've not heard it mentioned, but I'm pretty sure some (probably any US Corps of Engineers civil engineer) have wondered: "What is the prediction for these levees in the event of a thousand-year flood anywhere upstream of Reelfoot?" A reasonable prediction would be terrible consequences for people living in the floodplain, God forbid, but a happy event for the Reelfoot Lake ecosystem.

Spillways/Natural and Man-Made

All native lakes have outlets that control water levels in the lake: all lakes rise during wet seasons. Eventually, floodwaters fall and finally stabilize at the top of the lake's spillway; then it falls farther during droughts and dry seasons—nature's health spa, a way of treating part of the lake bottom to air and sunlight to rejuvenate its ecosystem.

The creation of a natural spillway is an interesting process. For example, when the river switches and leaves part of its old channel—does the old channel continue to function as a channel? Not exactly. Remember that these old segments become oxbows, sometimes called oxbow lakes, which happens when natural spillways are formed. The forming of a natural spillway has to do with the physics of sediments in flowing water. River water often carries a load of sediments (usually topsoil, alluvial sediments, from upstream erosion). As floodwater flows through the newly abandoned channel, it begins to slow in velocity as the flood subsides. When this happens, heavy sediments begin to precipitate and fall out, accumulate and build up in the old channel, more so at the outlet, which builds higher and higher to form a natural spillway. Eventually the inlet is built up as well, isolating the old channel—and an oxbow is formed.

The spillway at the creation of Reelfoot was a natural outlet. Eventually, a road was needed to get wagon teams across the outlet. The first road across the spillway was a corduroy log road. A levee and eighty-foot-long spillway was built in 1917, but was eventually considered too low. The construction of Highway 21 around 1920s raised the road (a levee) several feet, and a concrete spillway was constructed in 1937. The main purpose of this spillway was to stabilize the lake, hold the level of the lake for the convenience of establishing state property boundaries, of farming, boat docks, and other domestic uses of the lake. Nothing was mentioned about the needs of the lake's ecosystem, the future of fishing and hunting, outdoor recreation, economics, or any other thing. No one had much experience in wetland ecology or thought much about the history of the lake since hunting and fishing was the best in the state.

Artificial spillways, like the present Reelfoot spillways, are versions of natural spillways; the big difference is that we and not nature are ostensibly in charge (we are never fully in control of nature). So, the state and federal managers at the lake are charged with setting and implementing the policy for controlling lake stages. But nature abhors stability, and the practice of keeping lake levels stable at Reelfoot directly contradicted the principles of natural rivers and wetlands: river floodplains, including swamps and lakes, nearly always flood during spring and dry up during long, hot summers. The opposite had been the practice at Reelfoot now for more than forty years. Reelfoot lost its status as a natural lake in 1931, the year the Mississippi River mainline levee was built. Today, it is "semi-natural," and it is entirely up to us to tend it in order help preserve the natural features we now esteem as national treasures. We cannot recover the natural ecosystem once it is destroyed, but we can begin to duplicate many of its principles, enough to have reasonably, dependable and sustainable natural systems, like free-flowing rivers, natural hydrology that includes connecting wetlands to some periodic source of water, most likely the Mississippi River.

Lake managers recognized the contradiction. The extreme lake drawdown of the 1980s for Reelfoot was an effort to help get management of the lake partly back on track; managers had to be involved, now that the river was no longer the parent. We are. Reelfoot managers discovered rather quickly that wetland managers in Louisiana, Arkansas, and Florida (especially for major natural lakes and the Everglades) had come to similar conclusions. And we could take notice of lessons learned for Cato Lake, semi-natural lake at the Texas-Louisiana border.

The purpose of the drawdown at Reelfoot was to simulate dry seasons of summer. The counterpart was to allow lake levels to fluctuate a little higher when refilled to simulate the wet seasons. Dead and decaying plants and animals had accumulated several feet on the bottom of the lake. Nature's attempt to reduce this detritus to a healthy level required the lake bottom to be exposed to free oxygen (e.g. not bound in a molecule like H_2O)—the same free oxygen that fish and other aquatics need. One of nature's ways of solving this problem is to expose part of the lake's bottom to air and sunlight, which "burns up" dead organics and returns it to healthy soil that aquatic plants and animals require. Whether it could be done at Reelfoot Lake was a gamble and depended on two major things: the summer could not be rainy or the lake bottom would not sufficiently dry; and the spillway had to be capable of drawing down the lake quickly enough, deeply enough, and long enough to get the optimum effects of drying.

Good or bad, the second requirement failed: the old spillway was not capable of doing either—to draw down the lake fast enough or deeply enough to get the best affect. In addition, a federal lawsuit halted the entire project until an exhaustive study known as an environmental impact statement (EIS) was completed. Out of this study (which took ten years to complete) a new spillway with a capable design was proposed. Not until 2018 was the new spillway functional. Today, this new structure sits near the old one.

Part of the current policy is not to interfere with lake stages (to release water) during summer until the lake rises more than half a foot above "normal pool." During winter months, the spillway is not used to release floodwater until the lake is more than one foot above normal pool. The aim of the new policy is an attempt to mimic nature to some degree; that is, to allow lake levels to fluctuate, as nature would dictate, between floods and droughts or dry seasons. During normal years of native rivers and wetlands, their floodplains are usually flooded in spring and winter—the rainy season; their floodplains are dry during the summer—the dry season.

The proposed drawdown of 1985 will probably not be accomplished in the near future. By the time the EIS was completed, too much time had passed; old studies were out of date; allocated funds had been reallocated to other projects; and political and agency people supporting improvements at the lake were gone. Local user support for lake improvements today is unknown. I have mentioned the two major accomplishments for the drawdown proposal—a new and capable spillway and the purchase of buffer zones. The following are a few other things that came from the 1985 proposal: administration office and equipment storage facilities, a work base, a professional staff of managers and technical personnel, equipment to maintain roads and food crops for wildlife, hundreds of acres of reforested open spaces, and a planned strategy to deal with future needs of the lake.

The proposed drawdown during the mid-1980s was an enormous effort to improve the lake's ecosystem with better management capabilities. Much of that is represented in the development of the *Reelfoot Lake Fifty-Year Management Plan* and the *Reelfoot Lake Environmental Impact Study*. These studies have been helpful to identify critical management needs for this grand wetland, such as the design and construction of the new spillway, political and local support needed to management of the lake, and—the purchase of four thousand acres of buffer zones. Buffer zones have minimized landowner conflicts; allowed lake levels to be elevated and dropped as required without complaints; helped capture agriculture run-off of pesticides, phosphorus, and nitrogen, which kill or cripple tiny aquatic plants and animals that provide the foundation of the lake's food chain; and filter sediments to slow the filling of the lake from soil erosion.

In summary, the proposed drawdown provided an incentive to evaluate and provide solutions to numerous problems besides fish and wildlife habitat improvements at the lake, including the replacement of the old spillway. As mentioned, the new spillway is more efficient by being capable of controlling water levels needed in the future. The studies also identified management essentials of personnel, supplies, and equipment to clean the canals as well as the need to secure and develop new lands for wildlife. Along with the essential backing of local citizens and politicians, this period of support is the first comprehensive effort to address the problems of Reelfoot Lake since it became a public entity during the early 1900s. They should not be forgotten in the future management of the lake.

The Market Price for Native Wetlands

The mantra "Drain the Swamp" generally implies that someone might come and save us from mosquitoes, snakes, and swamp monsters—a putrefied swamp full of sickness and death that ought to be eliminated and made into dry farm ground. But Reelfoot Lake is a complex wetland, about half open lake and half in marshes and cypress swamps. But to some, it is only a swamp.

An interpretation of wetlands can be rather tricky when it comes to their protection and management, especially when 70 percent are on private lands. The subject of natural versus native wetland has not been well defined by anyone, let alone government agencies. Rather, well, all are just "wetlands" that fit under large categories. The US Army Corps of Engineers and the Environmental Protection Agency (EPA) have their own definition—a complicated one, which politicians catering to current economics keep changing. At this point in time, wetlands are essentially any land covered with water and has wetland vegetation. So a lousy wetland (man-made ones are usually a good example) deserves the same protection as viably vigorous wetland. This definition fails to distinguish between native wetlands and man-made wetlands—and that is foolish. There are paramount differences. Native wetlands are self-sustaining and have extremely valuable assets; man-made or artificial wetlands begin to deteriorate the day they are constructed, require constant upkeep, and generally cause far more detriments than benefits. Usually, this term refers to native wetlands converted into man-made wetlands. Channelized "rivers" and their wetlands are an example—instead of protection, they need restoration. Nearly all of them should be drained and the natural hydrology restored.

But there are several different kinds of man-made wetlands with different values. For example, fish culture ponds, farm ponds, and properly designed irrigation and waterfowl ponds can be valuable projects. The problem lies mainly with the definition of rivers: native rivers, converted native rivers, and man-made ditches. Like the definition of wetlands, the definition of a river remains a communication problem. The main point is that native rivers and their products are sustainable and immeasurably profitable; man-made streams (especially those that replace or substitute native rivers) have extremely limited benefits, if any at all, and are the antithesis of native river values.

A few major objections to man-made channels (channelization) constructed to replace native rivers include: (a) they soon degrade into hydrological and ecological chaos; (b) they kill thousands upon thousands of acres of valuable bottomland hardwoods from stagnant, ponded swamps and sediments; (c) they cause more flooding instead of less (the usual objective); (d) they obstruct navigation (channel depth is compromised); (e) they destroy nearly all of the fisheries and wildlife habitat; (f) they cause more soil erosion and transport excessive sediments and chemical pollutants; (g) they are a detriment to ground water supplies; and (h) they forever change the landscape in the interest of a few more acres of cropland (which is of questionable net success considering more land is flooded and much of it for a longer period). In a word, channelization of most native rivers and streams, and their wetlands, is ultimately a complete disaster designed primarily to accommodate one industry—agriculture. And the results of that objective are a net economic loss to the farming industry it is designed to protect, and the economics of the public purse. Every stream that enters Reelfoot Lake, and every West Tennessee tributary that enters the Mississippi has this characterization, including major stretches of the Hatchie and most of the Wolf River.

West Tennessee has thousands of acres of artificial or man-made swamps, mostly created by channelization (the construction of levees and ditches in lieu of natural rivers and streams). About the only positive value is that some hunters like these swamps for waterfowl hunting, which supersedes all other uses. Any hunter understands the trauma of giving up old hunting grounds, but there are better alternatives. Of course, to ignore these options implies their willingness to sacrifice the entire river floodplain of mature hardwoods, natural hydrology, fish, wildlife, and lumber values other hunters and landowners depend on. So what is my estimate of worth for these permanently flooded wetlands?

Less than zilch because they sacrifice many other valuable assets. Actually, every acre is ultimately a net detriment to society in general because, to save these man-made wetlands, requires the obstruction of the floodplain, and thus the destruction of the river's purpose to carry rain runoff and its ecosystem. This requires river restoration: wetlands that should be properly drained and the river restored; of course a natural swamp has an inestimable high value and should be preserved.

What then would be the value of Reelfoot? In a word, an inestimable value. Unfortunately, Reelfoot has been compromised. We must admit today that Reelfoot is a "semi-natural" wetland—only because it has been trapped by levees from its parent, the Mississippi River. And until recently, the lake's water levels were controlled by a man-made spillway, only for convenience. In other words, forethought about how water levels of the lake might affect the lake's biological health had not been considered.

So, I suppose, the question remains: What is the value of Reelfoot Lake? It can only be assessed completely once the lake is wiped off the map. It could be drained and placed into farmland. That might awaken us to the real value of Reelfoot. The entire society of businesses, parks, refuges, wildlife and fishery populations, a quarter-million annual visitor days, and a culture would be obliterated. The test of that proposal was made public in 1899–1908 and led to an awful uprising.

While the lake is not quite the same today, the potential value of the lost features enumerated above remains because the naturalness of Reelfoot Lake's ecosystem is still far greater than the man-made degradations caused since then. While economists might fudge an enormous value for the qualities of Reelfoot, they can never calculate the true value with money alone because meaningful dollar values for the intangible worth of social, psychological, or cultural entities—or the value of aesthetics and the simple presences of a natural ecosystem to future generations—cannot be grasped by economics. The flicker of light at the end of the tunnel is that we now have begun to understand what it takes to have sustainable, biologically healthy wetlands. But do we have the will and support to do it?

Sport fishing continues to hold the lead for outdoor use at Reelfoot Lake, and stimulates the greatest input to the local economy. It takes good quality water (free from the heavy runoff of pesticides, excessive phosphorus and nitrogen, and sediments that fill the lake) to have a healthy sport fishery—or any other outdoor recreation enterprise.

Filling the lake is not a new concern. Dr. Hal A. Baker, president of the West Tennessee Wildlife Federation in Jackson, Tennessee, expressed this concern to Conservation Commissioner J. Charles Poe in 1939. Not only did he recognize that levees had severed important ecological ties of the lake to the Mississippi River, but he also warned that erosion from cropland was filling the lake.[2] While the impact of his complaint is uncertain, an effort was made to retain eroded sediments by building small lakes in the watershed to capture sediments from eroded cropland. The agricultural industry has also made welcome strides in recent years to conserve topsoil and curtail soil erosion, with practices such as no-till farming, which requires herbicides. Herbicides are used to control weeds, and the practice disturbs the soil very little. Consequently, the public can be encouraged that sediments entering wetlands from soil erosion have been significantly reduced.

Still, many problems are at a critical point and far from being resolved. Agricultural chemicals are used on cropland around Reelfoot almost monthly and in large quantities. The chemicals used, and the effects these chemicals on the lake (and people), are unknown by the public, which remains skeptical. At present, the lake is poorly equipped to deal with the extent of harmful chemicals entering lake. Reelfoot is surrounded by an agricultural industry that depends heavily on the use of exposed soil and chemical treatment, which is reason to be concerned about the detrimental effects these chemicals have on Reelfoot wetlands. For one thing, common agriculture fertilizers of phosphorus and nitrogen have a great effect on microscopic plants and animals that become the basic energy of the food chain—too much or too little can cause a die-off and the collapse of the entire ecosystem.

So can the effects of pesticides. Microscopic aquatic plants and animals are very sensitive to insecticides and herbicides. One form of algae known as diatoms is ultra-critical to Reelfoot Lake; they recycle dead plants and animals (organic muck) for the food chain; they use up carbon dioxide to produce oxygen for aquatic life; they provide equilibrium to the ecosystem; and much more. These tiny phytoplankton are so critical to the biological well-being of Reelfoot, a die-off of this single species is enough to crash the sport fishery and the entire food web of the lake's fragile ecosystem.

Conflicts with agriculture, natural resources, and people's health cannot be ignored. Of course, agriculture is an essential industry for food. It can also very compatible with the environment—so long as it does not destroy more than it benefits. We need all agriculture and natural resource agencies on board to figure out how the industry can flourish while resolving these conflicts, especially at Reelfoot Lake. Politicians, the state health department, TWRA, Tennessee Department of Environment and

Figure 86. An aerial view of the Mississippi River and Reelfoot Lake
(background water): Mother and child.

Conservation (TDEC), and city and county officials should take notice
that the state is vulnerable to the loss of a multi-million dollar public and
private enterprise. Moreover, it stands to lose an irreplaceable natural
resource and public outdoor recreation area—unless the stewardship of
Reelfoot Lake meets the standards of a state natural area and resolutions
to protect it become practice.

The Stewardship of Reelfoot Lake

Reelfoot Lake is a designated state natural area. It was awarded National
Natural Landmark status in 1966 by the National Park Service, one of fourteen in Tennessee. It has recently been designated Tennessee's first National Heritage Site, and we wonder why it has not yet been registered in
the Ramar Convention record of critical international wetlands. Already
renowned as one of the nation's most unique natural resource treasures,
this recognition gives all the more reason why this finite and fragile lake
ecosystem should not be limited to common care but to special care. That
is, Reelfoot Lake should be placed high on the state's priority to assure
that state-of-the-art management is funded and implemented as a con-
tinuation program for the entire ecosystem—parks, outdoor education
and recreation, wildlife refuges, wildlife management areas, the fishery,

historical sites, and so forth—and the environment that influences this wetland and these programs.

Many programs have been proposed to preserve the integrity, use, and biological life of Reelfoot—sediment control; water level management; control or prevention of harmful chemicals such as fertilizer, pesticides, and herbicides; vegetated buffer zones to filter runoff and discourage encroachment; fishing, hunting, and boating regulations; conservation education through visual aids; miles of interpretative trails and circuitous boardwalks; reliable sources of water from river outlets to the lake; and the like. Some of these have been implemented to a degree; many have been proposed and forgotten. Proposals wither away in filing cabinets and end up in the dumpster when there are no leaders to keep them dusted off and implemented. All that I have mentioned needs dusting, updating, and follow-up.

The vigorous effort during the 1980s and 1990s to generate interest in the future management of Reelfoot Lake was long overdue but lot of headway was made. State and federal agencies, private citizens, and supportive political representatives tried very hard to find ways and means to do good things for the lake and its users. In fact, much was done—and the administrators, visitors, and local citizens should be very proud of these accomplishments. But that was almost forty years ago. Careers have been completed; the originators of the projects are gone—along with budgets, and citizen and political support. New managers are still implementing many of those programs, but they need help and encouragement to go ahead with new and better ideas and methods. But the momentum is on hold—on hold with much to be done.

Now is a good time for users of the lake and local interest to renew this effort with fresh faces, more knowledge, more talent, and better tools. State parks and wildlife agency administrators at the helm need a reminder. It would be useful to review some of the reports written by the main players of those years: the TDEC, Reelfoot Lake State Parks (RLSP), the Reelfoot National Wildlife Refuge (USFWS), the Tennessee Wildlife Resources Agency (TWRA), US Army Corps of Engineers (USCOE), UT Martin Biological Department, and others. The TWRA *Reelfoot Lake Fifty-Year Management Plan,* December 1986, is devoted in its entirety to the future management of the lake. Time, funds, and a lot of thought and effort went into these reports. These should be reviewed and revised to meet current conditions at the lake—and implemented as needed.

One of the issues will be how to incorporate the Mississippi River back into the lake. Before the mainline river levees, Reelfoot Lake received seasonable overflow from the Mississippi. In fact, old Civil War maps showing the system of Rebel fortifications and the operations of US Forces Oxbows under General John Pope clearly indicate that a natural channel still existed that ran from the river to the vicinity of Gray's Camp, just north of Tiptonville. A few years back, such connection with native wetlands prompted the USFWS out of Atlanta, Georgia, to lead a joint committee of scientists to explore ways to restore oxbows and other wetlands along the Lower Mississippi River Valley. The team met at the former Air Park Inn (including TWRA biologists, State Park managers, and Reelfoot Lake NWR managers). One of the suggestions was to consider a weir outlet from the river to the north end of the lake to assure a timely source of fish and water. The idea is very effective management for Lake Chicot, Arkansas, the largest oxbow on the Mississippi River. Other examples are lakes on the White River, and lessons can be learned from Cato Lake along the Louisiana-Texas border. Whether that program or one similar is appropriate for Reelfoot is yet to be determined. Nevertheless, this kind of effort will eventually find the nearest ideal management plan for the lake. However, it is not difficult to imagine that any suitable plan for Reelfoot Lake and its users would be too costly to implement.

While the intriguing and natural beauty of Reelfoot Lake will be with us a long, long time, it is a tired, middle-aged lake. It needs personal attention and first aid from top-notch land stewards—to be awakened, recovered and refreshed. America's unique oasis along the Mississippi has been constantly weighed down with too many thoughtless and careless designs and uses since our forefathers found it. What seemed our best efforts since it became Reelfoot Lake State Park 1917, we have hardly begun to address the most critical hurtles to provide the relief the lake needs.

Progress has been stalled long enough and needs leadership from the management agencies and public to move it along. Divided among three counties, two states, and three managing agencies, coordination of management policies and practices sometimes become cumbersome and difficult. The solution is to understand the problem, and then an agreement to fix it: non-partisan, non-political, practical, and for the benefit of both states and all Americans who treasure natural resources.

One place to begin is with our sister state Kentucky. A small part of Reelfoot lies in Kentucky, mostly federal refuge property. Among the difficulties is that Kentucky and its landowners have not been supportive partners, which has been a major stumbling block in the management of Reelfoot wetlands. Land management agencies cannot effectively manage land over which they have no authority. In this case, land managers in the Kentucky Department of Conservation are not authorized to participate in Reelfoot Lake projects in their state. It doesn't help that one US Senator from Kentucky repudiates the Department of the Interior's offer to purchase the conflicting wetlands that managers need to effectively manage those of Reelfoot Lake.

The conflict involves several hundred acres of cleared wetlands that lie north of the Reelfoot Long Point National Wildlife Refuge unit, most of it seasonally flooded farmland. This land is the Achilles heel of the total management of the lake. At the center of the issue is water level management; high lake levels spill into our sister state, and local Kentucky farmers and their politicians seem to have no outdoor recreation or conservation interest in Reelfoot Lake. It seems ironic that it has held up the options for the future of Reelfoot Lake and outdoor recreation benefits to Kentucky citizens as well as to Tennessee's. Not that managers and users do not empathize with these landowners, but there is a greater cause here—the survival of a twenty-four thousand—acre state and federal national treasure that affects millions of American citizens.

Absent political influence, the most apparent solution is for the federal government to enlist the State of Kentucky in a determined effort to resolve this longstanding problem. A common sense solution is to extend the refuge to include all Kentucky low land now in private ownership, considered below the historical high water mark. That response should make sense even to a US Senator. The new land would be primarily a buffer zone but managed under the same policies as the existing refuge. The lead agent must be the US Fish and Wildlife Service, Department of the Interior. Ironic also is that the Kentucky Department of Conservation during the studies of the 1980s and 90s saw this need and willingly attempted to participate. Once again, political interference nullified the progress. Purchasing land from willing sellers is not a new conservation effort. Tennessee has purchased from willing sellers about four thousand acres of low land around the lake as a buffer zone, mainly to minimize conflicts concerning high lake levels and landowner interests.

Mother and Child: An Irrevocable Relationship

The end of the Reelfoot Lake story has arrived. We know by now the great joy of using this valuable resource. We also know the need for accelerated professional management. And we understand the close natural relationship between the Mississippi River and Reelfoot is as a mother to a child.

This relationship between a river and its outlying wetlands is irrevoca-
ble—you can't have one without the other. Indeed, the tie is so strong that
both the primary ecosystem (e.g., the parent, Mississippi River) and the
subecosystems (e.g., the daughter, as the Obion River; or wetland, as Reelfoot Lake) will break down when separated and left to function on their
own. Daughter rivers and their wetlands will suffer ecologically the greatest, the parent less so. The parent river has many daughter rivers, and if
only one of the many is removed, it is probably be safe to say the daughter
will wither away like an artificial wetland; the parent will continue to survive—but disabled and less functional.

Since rivers stock and restore their temporary and permanent wetlands with aquatic life, nutrients, and rich soil, the entire ecosystem is
affected when the parent-child relationship is disturbed. In the short
term, it perhaps is more noticeable in oxbows and larger lakes because
the fishery will begin to deteriorate.

The American eel is a slender, four-foot fish often considered a snake,
but it is served worldwide as saltwater seafood. Seafood? It's a *catadro-
mous* fish, one that lives in freshwater and spawns in the sea. Some 80
percent of its former river habitat has been lost. We used to catch a lot of
eel on trotlines fishing for catfish. They are slick and slimy, an attribute
that allowed the fish to slip through small openings and escape a predator's grip of. Caught on a fish hook, they would somehow slime the trotline
for thirty yards in either direction. The eel needs freshwater habitat like
Reelfoot to complete its unusual life cycle.

Amazingly, the eel might live in the lake for twenty years before returning to the Mississippi River. From here, the eel swims fifteen hundred
miles from the Mississippi River to its spawning grounds in the Sargasso
Sea, a zone in the Atlantic surrounding Bermuda, where its migration
began. So the eel depends on a high-quality habitat, free from industrial
chemicals and obstructions on the way to the Sargasso Sea. I haven't seen
an eel at Reelfoot in many years. It should not go unnoticed.

Seasonal flow from the Mississippi was an important benefit to the
lake before the levees prevented it. Native wetlands severed from the river
become trapped when their free-flowing nature was forfeited, and they
soon withered down to artificial swamps, completely out of balance and
lacking the characteristics of robust fish and wildlife populations found
in natural floodplain lakes. The process is insidious and evolves over time.
One of the early symptoms is something I've already mentioned—"hyper-
eutrophic lakes" (a lake, swamp, marsh, etc., that resembles nutrient-rich

stable pools like farm ponds). Slow to recycle its nutrient loads, the ponded wetland no longer receives fresh water from the river to flush and renew it, and summer drying is not complete enough to fully decompose its accumulated dead plants and animals without free oxygen needed by fish and other aquatic life. Without completion of the drying process, these rich deposits fail to become a favorable habitat for the entire food chain, from tiny creatures to large predators. These were the same issues addressed by TWRA in a 1985 extreme lake drawdown proposal concerning water level fluctuations, although it was only partly implemented.

Quality and quantity of nesting and brooding habitat are essential for birds, especially migrant birds, whether these wetlands are for reproduction, or rest and food. All is forlorn if this habitat is unavailable, and they have no place to go when it's time to leave their brooding grounds. The "silent spring" of Rachel Carson's book on the detrimental effects of DDT might well be repeated if the quality and quantity of native wetlands fall below a certain threshold. In September 1919, the Cornell Lab of Ornithology (https://www.allabout birds.org) noted: "If you were alive in the year 1970, more than one in four birds in the U.S. and Canada has disappeared within your lifetime. Wild bird populations in the continental U.S. and Canada have declined by almost 30% since 1970 . . . every biome has dramatically declined in birdlife: the Eastern Forest Birds, the Arctic Tundra Birds, the Western Forest Birds, the Boreal Forest Birds, Shorebirds, and the Grassland Birds—2.9 billion birds, gone within 50 years."[3]

Where else do they have to go in the Mississippi River Valley but to the remnants of wetlands like Reelfoot Lake? Most of their habitat has been drained and cleared of wildlife. But we have overcome this crisis before and it can be done again—if we don't wait until it is too late. Look at the restoration success of white-tailed deer, wild turkey, giant Canada geese, otter, and other wildlife in Tennessee.

These wetlands are the "pearls" mentioned earlier, only the vestiges of wetlands here during my lifetime, usually mere patches of habitat found mostly in public parks and wildlife areas, scattered intermittently in river corridors. Some of these pearls are not far from Reelfoot Lake: Horseshoe Lake, Illinois; Ten-mile Pond, Missouri; and WMAs along the Mississippi River in Tennessee: Tumbleweed, Bogota, White Lake, Anderson-Tully, Shelby Forest, and Lower Hatchie NWR. This variety and quality of wetlands along the rivers and along the coastal plains assure that we will again see migrants on their return trip. Visible as oxbows, marshes, bottomland forests, and cypress swamps, these remnant wetlands lay as welcome mats for tired and migrating birds. More often than not, these are also the only remaining habitat for local wildlife. Reelfoot Lake is one of the most important of these remnant wetlands—for the austerity and integrity of the land, for fish and wildlife, and for people like us, who cannot be content without them.

We know Reelfoot Lake will be here for many more years as a very beneficial destination for outdoor users. Our challenge as users and managers is to maintain the quality as outstanding as our determination and skills allow. So, where will the managers go from here? They will keep on doing their job—hoping one day soon we'll all get together, do the right thing, and provide the stewardship needed for this rare oasis on the Mississippi.

Appendix 1

Other Books about Reelfoot Lake

Caldwell, Russell. *Reelfoot Lake: History, Duck Call Makers, Hunting Tales.* Self-published, 1988.

——. *Reelfoot Remembered.* Jackson, TN: Caldwell's Office Outfitters, 2005.

——. *Reelfoot Lake, the Tourist Guide.* Jackson, TN: Caldwell's Office Outfitters, 1989.

Clifton, Juanita (as told to Lou Harshaw). *Reelfoot and the New Madrid Quake.* Self-published, 1980.

Crockett, David. *Davy Crockett—His own Story: A Narrative of the Life of Davy Crockett of the State of Tennessee.* Carlisle, MA: Applewood Books, 1934.

Hayes, David G. *The Historic Reelfoot Lake Region: An Early History of the People and Places of Western Obion and Present Day Lake County.* Self-published, 2017.

Johnson, Jim W. *Rivers under Siege: The Troubled Saga of West Tennessee Wetlands.* Knoxville, TN: University of Tennessee Press, 2007.

Leonard, Lexie. *Reelfoot Lake Treasures.* Tiptonville, TN: Lake County Banner, 1991.

Orr, Arline Erwin. *"A Loving Community" called Phillippy.* Self-published, n.d.

Summer, Judy. *Walnut Log . . . A Place that Once Was.* Self-published, 1994.

Vanderwood, Paul S. *Night Riders of Reelfoot Lake.* Self-published, 1984.

Appendix 2

Evolution of Oxbows and Maturity
of the Floodplain

Recognizing the historical source of the unique mix of oxbow wetlands is yet another tip to help solve the mystery of Reelfoot's uncommon creation. Like all things in nature, the river never stays the same—it moves. Adjustments are made for better efficiency to carry raging floodwaters. Sometimes this shift occurs slowly, over years; at other times, suddenly, as during a single flood. Like all natural things, oxbows age and change over time, from one condition to another. First, sediments trap the abandoned bed, which might be a few miles (or several miles) long, and result in an open lake—like Chicot Lake in Arkansas. Sediments continue to accumulate over time, and the lake becomes shallower and smaller. Pioneering trees and shrubs like cypress trees, willows, and button bushes sprout and grow where the shallow shores dry long enough during the summer. Eventually, the oxbow is filled enough with sediments, and goes through various ecological stages to become a marsh surrounded by cypress swamps. Finally, the entire oxbow becomes land dry enough for trees to grow, as none sprouts and grows in water.

The entire stretch of land from near Hickman, Kentucky, to some forty miles south at the Obion River was low floodplain land that had many oxbow scars in various stages of natural succession. Along with small rivers, and a few bottomland hardwoods, this was the nature of the land beneath Reelfoot Lake before the early 1800s. Imagine what happened when the forest was permanently flooded! All dryland things were left as compost to decay. Like garden compost, all of this richness could not be useful to the garden until enough oxygen from the sun and fresh air was available to decompose the rotten material and turn it into biota-productive soil.

Bottomland hardwood forests are perfectly adapted to seasonal flooding and drying, the epitome of a healthy forest. Notice the living hardwood forests thriving in the floodplains of natural rivers; the floodplain is covered with water during wet seasons, but the water runs off as the river begins to fall, and the floodplain dries up. This is the secret formula that causes all natural wetlands to thrive; natural hydrology is the key for healthy rivers, lakes, and wetlands. Of course, this is remarkably absent in artificial wetlands and channelized rivers; the former hydrology is reduced to chaos; old waterways become obscure and nonfunctional; floodwater ceases to drain and thus becomes ponded; the rules of natural hydrology do not apply. Where rich land is permanently covered by ponded water, dead plants and animals that accumulate on the bottom of the pond struggle to decay. The only free oxygen available is that needed by fish and other aquatic animals. In moving waters, such as that of rivers and streams, oxygen is captured, replenished, and plentiful from tumbling water; in relatively stable ponds and lakes, oxygen can be suddenly used up by the respiration of plants and animals, or by dead plants and animals trying to decompose.

Unraveling the creation of Reelfoot is to have a greater understanding of the tremendous importance of river floodplains for the integrity of the country, its benefits to people, for outdoor recreation and conservation education, and for fish and wildlife. Within these scenarios are the principles needed to sustain native wetlands, and the potential to restore those we've lost. What the future holds for the continued stewardship of natural wetlands, or our best efforts to duplicate them, depends on how conscientious we are about conserving the planet, for in the wetlands we find the basics to a quality life.

It is helpful to be reminded that rivers are the creators of native wetlands—a mother and a child; a symbiotic relationship for the life of the wetland. The Mississippi River is an awesome subject, for it is the mother of wetlands for two-thirds of the nation; yet only with the one-time help of an earthquake was Reelfoot Lake created. The Mississippi Flyway is one of the most important corridors for migrant birds in the world. The importance of this flyway to America's bird life is so critical, its value is beyond expression. If we squander it, if we do not preserve it, we will face far more than the tragedy of silent springs.

Appendix 3

Wildlands to Domestic Lands

The lumber industry was a major source of income for residents during the early colonization of Reelfoot Lake. Bottomland hardwoods produce some of the highest valued lumber around the world, and the floodplains of the Mississippi River had the most and best—"the redwoods of the east." The supply must have seemed endless to the pioneers of the industrial age. Sustained harvest was not in the timberman's vocabulary back then, and needs considerable improvement today. A "forester" to most landowners was anyone trained to fell a tree, "experts" who nearly destroyed the forest and swrecked the land on the way out. Train loads of cypress and walnut logs, hundreds of years old, were shipped from Reelfoot Lake—some as a salvage operation when the forest in the basin of the lake flooded. Forest land from the river to the Chickasaw Bluff was cleared or heavily cut as a quick cash crop and to make room for cropland. It reduced wildlife land to a fringe around the lake, if any trees at all remained. Fence rows and property lines, normally left to mark these boundaries, were soon eliminated. Today, one can stand at the west bank of the lake and see the river mainline levee, changed from a wilderness to Kansas-style open space within two generations.

By the turn of the nineteenth century, very few large tracts of these trees were left in river floodplains—nearly all of it replaced mainly by short-season crops like soybeans. A few representative stands of these giant trees are still found around on the shores of the lake, and in bayous and state and federal refuges.

Even today, the modern mantra "Drain the Swamp" is deployed as a theme of economic progress. Our thinking has long been outdated. An attempt in 1917 to drain and farm as much of Reelfoot Lake as possible was a major issue in Lake and Obion counties. Citizens from wide reaches of the country came in protest. But the early 1900s were only the early stages of uneasiness from the general public that the next generations might have fewer and fewer wild lands to preserve. The need for the smell of fresh-turned soil, and the clearing of new ground in the back-forty was still in our pioneering blood; the mistaken smell of economic progress. Most thought we had plenty to spare then. But Master Plans with thoughtful trails to follow were not around back then, and very few contemplated for a hundred years hence.

Our land ethics are slow to change. By the 1960s, many soybean fields had encroached too far upon forested wetlands, and many acres were planted on land too wet to farm. In some instances, even the waters of Reelfoot Lake were held back by dams and the land farmed. Still, trees are an intrusion that flatland farmers despised. Not a square inch of land from ditch to ditch is spared in a thousand acres for even a meadow lark, let alone a bobwhite quail or cottontail rabbit. The thought of a "silent spring" on the farm has become a reality in most of the floodplains of the Mississippi River. Only the "despised" Snow Geese come to haunt them. Even today, wetlands are considered, for the most part, to be mosquito- and snake-infested swamps, better drained and planted in soybeans, or replaced by shopping malls, golf courses, and other things that match the priorities of our modern world.

Ignorant they are; without wetlands, our country will be sterile and devoid of huge populations of birds, and native fish and wildlife, outdoor recreation, runoff filtration, and more, much more than careless entrepreneurs care to know. Native wetlands will remain in a state of crisis until they are prioritized equally with development.

Notes

Chapter 1

1. See *New Madrid Earthquake Compendium Information of Eyewitness Accounts*, www.Memphis, edu; Vincent Nolte, *Fifty Years in Both Hemispheres* (London: Turbner & Co., 1854), 181–83.
2. See *New Madrid Earthquake Compendium*. A letter from James Fletcher to the *Pittsburg Gazette*, February 14, 1812, titled "Nashville, (Ten.) January 21 Earthquake."
3. Winifred L. Smith and Steve Pardue, *Interpretive Research Document for the Reelfoot Lake Interpretive Center* (Tennessee Department of Conservation, 1985), oxbows and, 39; the formation of Reelfoot Lake and, 44. Smith is a research limnologist, historian, and professor emeritus at the University of Tennessee at Martin, Tennessee. Pardue was the Reelfoot State Park superintendent during this period. This work cannot be surpassed as a complete narrative summary of the settlement, folk life, description, and general history of Reelfoot Lake supporting the topics and displays of the Reelfoot Lake Donaldson Memorial Museum.
4. Enric Sala, "The Cost of Harming Nature," *National Geographic Magazine* 238, no. 3 (September 2020): 15–18.
5. Mark Twain, *Life on the Mississippi River* (Seawolf Press, 2018), 1, 2.
6. R. C. Donaldson, "Pioneering Life in Madrid Bend (Ky.)," July 25, 1947, *Lake County Banner*, November 4, 2020, 2.
7. The Mississippi River Commission, "Lower Mississippi River Early Stream Channels (A drawing)," Vicksburg, MI, 1938.

Chapter 2

1. *Lake County Banner*, September 23, 2020, 2; an editorial by R. C. Donaldson, "How the Washout was Formed," May 30, 1947.

Chapter 3

1. Glenn Hodges, "Cahokia," *National Geographic Magazine*, January 2011, 127.
2. Ibid., 145.
3. David Crockett, *A Narrative of the Life of Davy Crockett of Tennessee* (Bedford, MA: Applewood Books, 1834), 89.

4. R. C. Donaldson, "Dawn on Reelfoot," July 18, 1947, an editorial under "Lake County Bygones," *Lake County Banner*, October 2020, 2.

5. David G. Hayes, *The Historic Reelfoot Lake Region – An Early History of the People and Places of Western Obion and Present Day Lake County* (Instant Publisher.com, 2017), 127, 128.

6. R. C. Donaldson, "A National Highway to the West," November 21, 1947, an editorial under Lake County Bygones, *Lake County Banner*, January 13, 2021, 2.

7. Crockett, *Narrative of the Life*, 103.

8. Ibid.

9. Constance Rourke, *Davy Crockett* (New York: Harcourt, Burke and Company 1934), 86–87.

10. Crockett, *Narrative of the Life*, 108.

11. Ibid., 16–17

12. Hayes, *Historic Reelfoot Lake Region*, 32.

Chapter 4

1. Hayes, *Historic Reelfoot Lake Region*, 31–35.

2. Crockett, *Narrative of the Life*, 108.

3. R. E. Lee Eagle, *Reelfoot Lake Fishing and Duck Hunting* (Nashville, TN: McQuiddy Printing Co. 1915), 4.

4. Lexie Leonard, *Reelfoot Lake Treasures* (Tiptonville, TN: Lake County Banner, 1991).

Chapter 5

1. Rourke, *Davy Crockett*, 86–87.

2. Ibid.

3. Brown Johnson, et al., *Reelfoot Lake 50-Year Management Plan* (Tennessee Wildlife Resources Agency, 1988), 74–75.

Chapter 7

1. Infrared aerial photograph of Reelfoot Lake and vicinity. This photograph has been in the author's files since his employment with the Tennessee Wildlife Resources Agency (retired 2005). It is probably a duplicate USGS or an SCS photograph, 1980s.

Chapter 8

1. E. O. Gersbacher, "The Heron Rookeries at Reelfoot Lake," *Journal of the Tennessee Academy of Science 14* (1939), 162–80.

Chapter 9

1. Don Marquis, *The Washington Herald* (Washington District of Columbia), January 10, 1909, 24.
2. Hayes, *Historic Reelfoot Lake*, 310–13.
3. Paul J. Vanderwood, *Night Riders of Reelfoot Lake* (Tuscaloosa: University of Alabama Press), 146–47.
4. Note: Newspapers of that period across the country were filled with accounts of the 1908 Night Rider murders and trials. In general, the citations above capture or summarize most of these accounts, the court records, and interviews with members of the clan and local people.

Chapter 10

1. Twain, *Life on the Mississippi*, 1–2.
2. http://1Mississippi.org. 1Mississippi is a citizens' action and education organization dedicated to the wise management and use of the Mississippi River and the preservation of its natural resources.

Chapter 15

1. Twain, *Life on the Mississippi*, 219.
2. Lydel Sims, "Letter to Poe Suggests Steps for Filling in by Silt Deposits, Vegetation," *The Jackson Sun*, Jackson, Tennessee, April 12, 1939, 8.
3. Gustave Axelson, *All About Birds*, https://www.allaboutbirds.org/. The Cornell Lab of Ornithology, "Vanishing: More Than 1 in 4 Birds Has Disappeared in the Last 50 Years," September 19, 2019, http://www.birdscornell.edu/?.

Index

Air Park, 102–3, 148, 198, 220
airboats, 108, 153
American egret, 105, 171
amphibians, 8, 136, 190, 191, 202
Asian carp, 60, 77, 184, 186, 188
avulsion, 8, 10, 20

Baker, Hal A., 216
bald eagles, 8, 102, 130, 147, 150; eagles
    and osprey nesting, 154–57, 160,
    161, 182, 183, 192, 193, 204
Barker, Sherrill, 181
Barkley Lake, 12, 13
Basham, "Red," 105
bass fishermen, 186, 187
Bayou du Chein, 11, 20, 27, 33, 34, 40;
    former river, 51, 52, 69–92, 111, 116,
    117, 176, 177, 199
Bennett Johnson Road, 72, 88
Berry, "Boochie," 60, 67
bird watchers, 194
birders, 130, 147, 150; Reelfoot as a
    birder paradise, 193–96
Black Bayou Refuge, 16; as a model for
    managing wetlands, 99, 101, 121,
    131, 132, 147, 172, 175, 197
Blue Basin Lodge, 157, 198
Blue Wing Club, 54, 210
Blue Wing Club House, 54
bluegill fishing, and spawning season,
    80, 159, 188
boardwalks, as access across the
    bayou, 55, 57, 77–80, 102, 110, 125,
    148, 158, 171, 172, 194, 198, 201, 219
bobwhite quail, 162, 168, 169, 230, 231

Bo's Landing, 204
Bratten, Lonnie, 106
buffalo, 59, 60–62, 85, 92, 106, 184
buffer zones, 98, 131, 132, 200, 230, 219
Burdick, J. C., Jr., 33, 38, 39

Cahokia, 25
Canada geese, 123, 126, 127, 180
carp, 13, 14, 59, 60, 63, 77, 107, 184, 186,
    188
Campbell, George, 35
campgrounds, 102, 176
Carter, James, 71, 94
Carringan, Helen, 59
Cates Landing, 169
catfish, 59–61, 80, 83–85, 92, 106, 111,
    112, 171, 172, 176, 183–85; and fish-
    ing techniques, 188–90, 222
Champey Pocket, 16, 34, 45, 197
channelization, 98, 207, 215
Chickasaw, 6, 12, 16, 25–28, 30, 38, 40,
    42, 45, 50, 75, 89, 102
Chickasaw Bluff, 26, 40, 42, 45, 75, 89,
    103, 173, 200, 205
Chicot Lake, artificially managed
    oxbow, similar to Reelfoot Lake,
    229, 227
Civil War, 35, 38, 44, 54, 204
Civilian Conservation Corp, 79, 204
commercial fishing, 38, 41, 48, 51; as a
    fading source of livelihood, 57–62;
    67, 72, 80
cookie cutter, 148
cotton mouth snake, 71, 86, 87, 117; as
    poisonous, 190, 191

Crane Town, 172
crappie, 58–61, 92, 149, 151, 152, 159, 183, 184, 187, 188
Crockett, David, 26, 28–31, 35–39, 50, 65
Crossley, Lillian, 57
Crossley, Steven, 51, 59
crows, 164
cypress trees, 10, 20, 23, 28, 38, 52, 74, 81, 100, 108, 171–73, 181, 201, 227
Cypress Point, 178, 201, 204

DD-T, 155
Dehart Store, 93
de-snagging, 149
Donaldson, R. C., 28, 164
Donaldson Canal, 149
drainage districts, 43
drawdown, 203, 204, 212, 213, 223
duck blinds, 56, 135–45, 174, 180
ducks, 41, 113; and fall migrations along the Mississippi Flyway, 120–22, 124, 126, 127; and hunters, blinds, and sportsmanship, 136–44, 193

eagles, 130, 147, 154–58, 182, 183, 192, 195, 197
ecology, 1–3, 7, 12, 23, 67, 110, 127, 211
Ellington Center, 102, 192, 194, 204
exotic fish, 184, 186, 188

Fish Gap Hill, 173, 174
floods, 9, 16, 22; as the 1919 "River-Reelfoot Lake-Levee-Road Project," 40–45; and living with, 56–58, 74, 76, 77, 173, 203, 205, 207–10, 213
fly fishing, 188
"flying turtles," 156
forests, 20, 30, 57, 63–66, 131, 132, 226
forester tern, 160
freeze-ups, 181, 183
frogs: bull frogs, 46, 77, 84–87, 117, 160, 161, 191; and cricket, 71, 84, 85,

191, 192; and green tree, 13, 160, 191, 192; and kinds of, 191, 192; and leopard frogs, 191

Game Warden Shack, 189
giant Canada geese, 8, 126, 127, 223
giant cutgrass, 70, 77, 117, 122, 141, 142; as falsely considered the "Curse of Reelfoot Lake," 152–54, 195
"Godfather, the," 49, 58
Golden Age, 53, 54, 56, 67, 118
Grassy Island, 26, 54, 79, 91, 95, 106, 132, 148, 158, 170, 200
Gray's Camp, 33–35, 39, 45, 51, 85, 91, 137, 148, 166, 198
Gray's Camp Lodge, 178, 198, 200
great blue heron, 80 105, 128
gumboots, 40, 46
Gumbooters, 45–67

Hamilton's Camp, 178
Harris, J. C., 62, 14, 115, 118
Harris, Judge, 62, 114, 115, 118
Hatcher, Bob, 155
Hayes, David G., 118
Hayes, Marvin, 58, 59
hiking trails, 152, 158, 171
Horse Island Canal, 153, 182
hyper-eutrophication, 202, 222

Isom Lake refuges, 132

Jackson Purchase, 22, 38
jet spray, 148
Johnson, Bennett "Ras," 47, 83, 99
Johnson, Mildred, 47
Johnson, Willey "Pappy," 137

Kentucky Bend, 10, 12, 23, 34, 38, 39, 42, 168, 169, 170
Key Corner, survey post for the Western District, 27

Keystone Lumber Company, 66
Kirby Pocket, 175, 204

Lake People, 37, 40, 43, 44, 50, 76
Leonard, Lexie, 3, 36, 67
levees: built along the Mississippi
    River to protect farmland, 105–8,
    210, 215–17; districts, 42, 43; faith
    in, 75, 76, 100–103; part creator of
    Reelfoot Lake, 8 16
Little Ronaldson, 65, 105, 172
Lovell, Tommy, 138
Lower Blue Basin, 185, 186
Lyons, Alfred, 51

manmade wetlands, versus natural
    wetlands, 131
Mark Twain, 38, 71, 123
mayflies, 77, 188
McQueen, Bruce and Annie, 67, 176–
    79; and Charlie and Lucy, 28; and
    Grover, 74; and Harry, 77; and
    James, 106, 147; Lula, 83; and
    Tobe, 71
McQueen's Grocery, 52, 55, 70, 73, 112
Miller's Camp, 178
Miller, Don, biologist, built osprey nest-
    ing platforms, 155
"mine fields," 20, 21
Mississippi Flyway, 119–33
Mississippi River, as mother to a child,
    218, 221, 228
Morris, Wendell, 63, 67, 106
mud snake, 132, 199
Myers, Gary, as director of TWRA, 97,
    99

National Heritage Site, 218
National Natural Landmark, 218
Nation's Camp, 178
Native Americans, 2, 25–27
New Madrid, 9–12, 16–18, 19, 27, 40, 167
Night Riders, 88, 115–18

Night Riders of Reelfoot Lake (Vander-
    wood), 118
Nolte, Vincent, 5, 6

osprey: arrival in spring, 147; nesting
    osprey and eagles, 150–57, 160, 192
oxbows, 205, 211, 220, 222, 223, 227, 228

Parsons, Herb, 139
Pelican Fest, 103, 176
Pitts, David, 129
Poe, J. Charles, 116, 117
Pole Road, 28, 29
pontoon trips, 152
Powell, Spencer, 71, 84; and Myrtle, 88,
    116; and Verge, 81

quail, bobwhite, 131, 162, 168, 169, 230

Rankin, Capt. Quentin, attorney hung
    by Night Riders, 116
Rankin, Luther, 118
red bobber, 187
"Red-foot Lake," 31, 35
redstart, 195
Reelfoot Commercial Fish Co., 118
Reelfoot Creek, 25, 27, 31, 95
Reelfoot Lake State Park (RLSP), 102
Reelfoot National Wildlife Refuge
    (RNWR), 95, 219
Reelfoot River, 27, 40, 65, 91
reptiles, 136, 147, 154, 160, 183; about a
    dozen different kinds, 190, 191, 202
Robinsons, 176
rookeries, 105, 130, 170, 171, 193
"rough fish," 59–61, 183, 185
Round House Park, 204
Rourke, Constance, 30, 37, 38
Rutherford, Henry, 26–29, 31

Samburg, 38; first known as Wheeling,
    40 42, 45, 51, 53, 59, 66, 75, 91, 113,
    114, 136, 148, 178, 197, 200, 201

sand piper, 158, 195
sandhill cranes, 128
sawmills, 42, 63, 65
"Scatters," the, 41, 43, 92, 96, 97
Scheland, George, 8; and James, 106, 147; and Lula, 83
Shaw, John, 116
Sherwoods, 176
"Slough," the, 52
snapping turtle, 83, 86, 88–91, 163
snow geese, 180, 204, 230
"Son of the Swamps," 56, 94
Spicer, Elbert, 140, 142
Spillway (or spillway), name of structure, 201–4, 210, 214
sport fishing, 147, 159, 202, 216
"sports," 45, 49, 51, 73
Sportsman's Resort, 198
Spot, David and the boy's dog, 71
State Woods, 100, 128, 197
steamboats, 37, 38, 39, 123
Strader, Onice, 194
"string of pearls,"Mississippi River islands, 124, 130, 228
stump-jumper, 17, 46, 55, 66, 67; custom made for Reelfoot, 70, 73, 81, 89, 92, 93
sunsets, 200
Swamp People, the, 40

Taylor, "Slingshot" Charlie, 142
Tennessee Wildlife Resources Agency (TWRA), 124, 125, 145, 216, 219
Tiptonville Dome, 19, 45
trammlers, 106
tree swallows, 157, 158
turtles, 71, 73, 83, 86, 89–91, 147, 156, 161–64

U.S. Marshals (film), 105, 106

Vanderwood, Paul J. (Night Riders of Reelfoot Lake), 118

Wallaston, O. T., 51, 59
Walnut Log Lodge, 52, 59, 73–74, 78
Ward's Hotel, 116
Washout, 16, 17, 50
"weeds," 131, 133, 163, 168, 189, 202
West Tennessee Land Company, formed by Judge Harris to possess ownership of Reelfoot, 114–16, 118
water snake, 191
water willow, 77, 153, 154
woolies, 188

Young, H. B. "Con," 116
yo-yos, 176, 189